D0487627

Oxford Science Publications
COSMIC LANDSCAPE

COSMIC LANDSCAPE

Voyages back along the photon's track

MICHAEL ROWAN-ROBINSON

Oxford University Press · 1979

Oxford University Press, Walton Street, Oxford OX2 6DP

OXFORD   LONDON   GLASGOW
NEW YORK   TORONTO   MELBOURNE   WELLINGTON
IBADAN   NAIROBI   DAR ES SALAAM   LUSAKA   CAPE TOWN
KUALA LUMPUR   SINGAPORE   JAKARTA   HONG KONG   TOKYO
DELHI   BOMBAY   CALCUTTA   MADRAS   KARACHI

British Library Cataloguing in Publication Data

Rowan-Robinson, Michael
   Cosmic landscape.
   1. Astronomy
   I. Title
   523   QB43.2   79-41327

   ISBN 0-19-857553-X

Typeset by Oxprint Ltd., Oxford
Printed in Great Britain
by Billing & Sons Ltd., Guildford and Worcester

To Mary, Adam, and Jonathan

PREFACE

In *Landscape in art*, Kenneth Clark says that things that can only be seen with a telescope cannot possibly be part of our landscape. From my own experience of the past fifteen years, while I have been working in astronomy, I think that this is not necessarily true. The stars and galaxies can by an act of imagination become as real a part of the landscape as the trees and hills of the terrestial foreground.

I remember that when I started to do research in astronomy, I found it a disconcerting pursuit at first. I had to learn an entirely new scale of distances, abandoning inches and miles for light-years and millions of light-years. I had to cut myself off from my own perceptions and common sense in order to grasp fully this new universe.

In recent years I have found myself returning again and again to earth and the human scale of things. While writing this book I kept finding the solar system and earth itself looming larger and larger. I have tried to keep my eyes on the stars and galaxies, while at the same time holding the trees and hills, the beaches and oceans of the foreground in focus, like the illustrations of some of those wonderful nineteenth-century popular books on science.

One of the most dramatic changes in our perception of the cosmic landscape has come in the past few decades, with the opening up for astronomy of the invisible wavelengths: radio, microwave, infrared, ultraviolet, X-ray and gamma ray. Suddenly there are a whole series of entirely new landscapes to be visualized. This experience can only be compared with Galileo's first days with the telescope early in 1610: almost night by night questions that had perplexed people for thousands of years were answered. Galileo vividly described this opening up of our horizon in his pamphlet 'The Starry Messenger'. The opening up of the invisible wavelengths for astronomy has been the work of thousands of men and women, and much of their story remains to be written. Although many popular books on astronomy have appeared in recent years, the landscapes of the invisible wavelengths remain inaccessible to most people.

In this book I try to share this experience of the new astronomy with you. I have tried to write the book for someone who might never have read anything about astronomy, but I realize that some of the ideas described here will seem strange and difficult. In order to try to help the reader make the imaginative leap needed to see the cosmic landscape, I have, rather than simply paint the landscape as it appears from earth, used the image of a series of voyages in each of the main wavelength bands of contemporary astronomy. In each band we travel back in time along the path that the light has taken to reach earth. Since the things we want to look at are in different directions on the sky, these voyages do not correspond to a route that any sensible space navigator would take. They are more like a succession of views in a telescope with different magnifications. However I sometimes find myself taking this metaphor of a voyage literally and describing how I think things would look from a different angle to that seen from earth or, say, looking back to see what earth looks like.

Certain objects in the cosmic landscape figure in several different wavelength bands and so our picture of them builds up only gradually. Some readers may find this a limitation of my approach, that they have to wait till later voyages to find out the meaning of what they see on an earlier one. However this does correspond rather well with how astronomy develops, with new wavelengths and new techniques providing the clues that were missing before.

Several colleagues at the University of California, Berkeley, very kindly read parts of the book and saved me from numerous blunders: Stuart Bowyer, Martin Cohen, Joe Silk, and Jack Welch. But I didn't heed all their warnings and the surviving errors and excesses are all my own. Michael Rodgers has been a constant source of encouragement and enthusiasm. His hundreds of comments on the first draft of the book were invaluable. And my wife, Mary, struggled to keep the book, to use the current California dialect, grounded. Without her efforts the book would probably have been impenetrable to the general reader.

Kensington
California
April 1979   M. R.-R.

CONTENTS

1
INTRODUCTION

The meaning of landscape

Until it is tamed, landscape can seem harsh and hostile to human beings. Parts of the earth are permanently hostile by virtue of extremes of temperature — the polar wastes and the tropical deserts. Elsewhere people live with the shadow of earthquakes, volcanoes, floods, and hurricanes continually hanging over them. Even in more placid environments the farmer breaks his back over the soil and curses the fluctuations in rainfall which ruin his crops.

It is not surprising that so many images of nature, among both primitive and civilized peoples, are sombre and cataclysmic. Hence also the attraction of the closed garden, the secure, tamed piece of nature walled off from the world. Here man can gaze on nature in a calm and serene, but escapist, frame of mind.

In Western civilization it is during the Renaissance that we first find a more positive attitude to nature. The first person to climb a mountain to enjoy the view seems to have been Petrarch, and in his writings there is an all-embracing love of nature and the countryside: 'Would that you could know with what joy I wander free and alone among the mountains, forests and streams.' We find for the first time the concept of landscape, nature experienced as a whole by an individual: nature as a source of wonder and joy.

This all-embracing admiration of nature is at the heart of the scientific view of the world. It is not surprising that it should find its expression, however, in poetry rather than in the writings of the scientists themselves. The scientist is trying to comprehend the richness of nature and its phenomena rather than to convey this richness to others. Even so some scientific books, like Galileo's *Starry messenger* or Darwin's *Origin of species*, have become classics.

The earth offers such an immense range of landscape, from the dramatic wilds of desert and canyon to the patchwork of hedge

1

and field of England, where even the remotest fell or moor bears traces of human occupation. But for me the essence of landscape is that there is a human being at the centre who experiences it. This is the sense in which I shall speak of the cosmic landscape. The landscape is an individual and personal experience. For example, each of us carries in our mind the landscape of childhood, those first recallable images of the outside world. And whether this landscape be harsh and alien or a secure garden it remains vivid and important throughout our lives.

The cosmic landscape

When we look up at the night sky and see the moon, the planets, the stars, they do not seem part of our individual landscape. They are so remote and ethereal. We cannot travel to them, around them, we cannot look at them from different angles and from different distances as we can with the the terrestial landscape.

We see so little detail. To the naked eye only the moon shows any features. (To the inhabitants of ancient Teotihuacan in Mexico these features resembled a rabbit and they had an improbable story of how it got there.) The planets are slightly extended and they show some faint colour, as do the brighter stars in clear conditions. The changes that occur — the phases of the moon, the motion of the moon and planets through the stars, the nightly rotation of the sky, and the change from summer to winter constellations and back — all these have to be watched for regularly to see any pattern. Modern man is probably on average more ignorant of the night sky than the agricultural peoples of the remote past who used the aspect of the night sky to tell the season accurately. Indeed we have to marvel at the astronomical achievements of earlier, pre-telescopic cultures: the patient observations of the ancient Chaldeans in discovering the 18 years and 11 days cycle in the time that eclipses of the sun and moon occur, the care with which Chinese astronomers of the first millenium AD surveyed the stars to find changes in brightness or new visible objects, the precision with which the Maya were following the orbit of Venus at the time of the destruction of their books by the zealot Bishop Landa in 1562, the subtle discoveries of the medieval Arab astronomers, and the monumental work of the Greeks in the first 500 years BC.

There is a fascinating account by the Jesuit Lecomte of the working of the Chinese Astronomical Bureau as he found it in AD 1696:

They still continue their observation. Five Mathematicians spend every night on the Tower in watching what passes overhead; one is gazing towards the Zenith, another to the East, a third to the West, the fourth turns his eyes Southwards, and a fifth Northwards, that nothing of what happens in the four Corners of the World may scape their diligent Observation.

It is very moving to think of the five mathematicians on the Tower, night after night for fifteen hundred years.

Of course an important change in our awareness of the cosmic landscape has happened in our times. Through the eye of the television camera we have ourselves seen the surfaces of the moon and Mars. I remember vividly the meeting of the Royal Astronomical Society in London in 1966 at which a slide was shown of the first picture of the moon's surface from NASA's *Surveyor I* camera. Suddenly there was a landscape with rocks and hills, rather like one of Leonardo's rugged backgrounds. Later there were the astronauts lumbering about in slow motion. And even those inarticulate pioneers conveyed one powerful image: how beautiful a haven the earth looked from that hostile terrain. The whole earth becomes the enclosed and secure garden, protected by the walls of its atmosphere and magnetic field. And the moon and Mars, scenes of so many wonderful fantasies of life on other worlds, are added to the list of magnificent but desolate products of attrition, the canyons and deserts. Places for the adventurous or foolhardy to visit in person but for all to visit in imagination.

How can this cosmic landscape of stars and those gigantic star systems, galaxies, become real for us? We have to travel in imagination to the universe's furthest shore. The cameras of the planetary probes take us only a minute step from our own soil. To comprehend the universe of the modern astronomer, we have to abandon our human scale of measurement. We have to leave solid ground behind, since lumps of rock like the earth are so rare. We have to become explorers without preconceptions.

The voyages

From the fifteenth century onward the horizon of the European world began to be enlarged by the voyages of the navigators and explorers. After the voyages of Vasco da Gama, Columbus, Amerigo Vespucci, Marco Polo, and Magellan, new continents, new peoples, new plants and creatures swam into the popular consciousness.

Later came the scientific voyages of the *Beagle,* on which Darwin came to understand the origin of species, and of the *Challenger,* which revealed the secrets of the ocean depths. To fill in the blank areas on the map of Africa became the goal of the nineteenth-century adventurers, leaving only those grim treks to the poles early this century to complete the exploration of the globe.

These voyages of discovery remain a powerful image of human development and aspiration. The intensity of this image arises not only because the voyages represent in a striking way mankind's cultural evolution, the progress of his collective knowledge, achievement and consciousness. It is also not just that they provide colourful illustrations of the Aristotelian virtues of courage and fortitude. The point is that they symbolize the individual psychological development that each of us goes through, the enormous expansion of horizons that we experience in childhood and adolescence.

What are the modern equivalents of the voyages of discovery? At first sight it seems to be the space programme, the landing of men on the moon, and the exploration by television camera and other scientific instruments of Mars, Venus, Mercury, and now Jupiter and Saturn. As if to mimic Columbus or Captain Cook, the astronauts step out, plant a flag, say a prayer, pick up a handful of soil, and so on. Their planners and navigators, safely back in the control room at Houston, are certainly conscious of history.

In reality the most spectacular opening of horizons in these decades has come from the great telescopes and the opening up for astronomy of the 'invisible' wavelengths: radio, ultraviolet, X-rays, infrared, microwave. The years since the Second World War have been the Golden Age of astronomy, an experience comparable in its totality only with Galileo's first sight of the night sky through a telescope. That at least is the case I shall try to make in this book.

What is light?

What is light? It seems to bathe our whole consciousness. We describe our most profound experiences as 'visions', 'illuminations'. It seems so fundamental to us, we are not aware that we have to process the waves entering our eyes in order to extract information about colour, brightness, and shape. The whole picture seems to present itself to us, complete, in an instant. To get some idea of the nature of light it's better to look at some detector less versatile and brilliant than the human eye, one that is not bound into our thinking.

Let's look instead at the domestic radio receiver. But surely, you may say, radio waves are not light? It may come as a shock to learn that the waves entering your eye and those entering your radio are identical except in one respect. Their colour. Or to use more exact terms, their wavelength, their frequency. Radio waves are light of a frequency (wavelength, colour) that the human eye cannot see.

It is quite difficult to get an idea of what kind of waves light consists, so we'll leave that for the moment and return to it at the start of our radio voyage. We'll just say that light consists of waves travelling through space. The wave-like properties of light are rather subtle and were only discovered at the turn of the nineteenth century, through ingenious experiments. However you can do a very simple experiment to see something caused by light's wave-like nature. Hold your first finger and thumb very close together, almost touching, near your eye and look into the distance. You will see dark lines echoing the outline of your finger and thumb which are caused by the light waves interfering with each other and cancelling out. The nature of the waves, i.e. what it is that is waving about, was not understood until the classic work of the Scot, James Clerk Maxwell, on electromagnetism in the middle of the nineteenth century.

Think instead of the waves on the ocean, travelling towards the shore. It is obvious that it is not the whole ocean that is travelling to the shore, but only a disturbance which is propagated along the top of the sea. The water at any spot moves in a circular path as the wave passes, but there is no net motion of the water, only of the wave. As the crest of the wave passes, a drop of water in the surface of the ocean is moving forward with the wave; as the trough passes, the drop is moving backwards. As you stand at the shore watching the waves break it is basically the same water surging up and down the beach all the time, apart from the effects of local ocean currents. The waves arrive at the shore every ten seconds or so: we say the frequency of the wave is one per ten seconds, or 0.1 per second. This unit of frequency, the number of cycles per second, has been given the name of the hertz, after the nineteenth-century German physicist, Heinrich Hertz. We can say our ocean has a frequency of 0.1 hertz.

Now the spacing between the crests of the waves, or for that matter between the troughs, is about 10 metres. This spacing is called the wavelength. The speed with which the wave advances the ocean is therefore about 10 metres every 10 seconds or 1 metre per second. This relation, the frequency multiplied by the wavelength is equal to the speed of the wave, holds for all types of wave. The speed of waves on water depends on the depth of the water and on the type of wave. Waves on the surface of the ocean, generated by the wind blowing across it, move faster as the wave approaches the shore. Tsunamis, the 'tidal' waves caused by underwater earthquakes, move much faster than surface waves.

The speed of light also depends on what kind of medium it is passing through — it is slightly slower for light passing through water than through air, but there is not much difference between the speed of light through air and through a vacuum. Most of the universe is very empty, a much better vacuum than we can make on earth. The speed of light through a vacuum is 300 000 kilometres per second, and this is so fast that it seems instantaneous in our own local experience. But when NASA ground control talked to the astronauts on the moon, there was always that curious three-seconds gap between question and answer, as if the astronauts had to scratch their heads to answer the simplest query. In fact this is the time the radio signals took to travel to the moon and back. Radio waves travel at the same speed as visible light, and indeed as infrared waves or X-rays – and all the kinds of light we shall be talking about in this book. So light from the moon takes 1.5 seconds to reach earth.

This gives us a new way of measuring distance, by the time it takes light to traverse it. We can say the moon is 1.5 light-seconds from earth. The light from the sun takes 8 minutes to reach earth so it is 8 light-minutes away. And the light from the nearest star takes 4 years to arrive, so this is 4 light-years away. A light-year is ten million million kilometres and once we travel out into the landscape of stars and galaxies it is best to forget about kilometres and try to think in light-years. The furthest stars visible to the naked eye are several thousand light-years away and the light from the most distant objects known to us in the universe has taken more than ten thousand million years to reach us. Light which crosses our room in a fraction of an instant seems to traverse the universe at a snail's pace, so vast are the distances involved.

Let's go back to our radio receiver. It usually has several different bands available: 'long wave', 'medium wave', or 'very high frequency' (VHF). The long waveband consists of wavelengths in the range of 1000–2000 metres, or frequencies 150–300 kilohertz (1 kilohertz, or kHz, = 1000 hertz = 1000 cycles per second), the medium waveband is wavelengths 200–600 metres, or frequencies 500–1600 kHz. The VHF band is 3–3.5 metres or 88–104 megahertz (1 megahertz = 1 million hertz). It's odd that the long, medium, and short wavebands are named because of their relative wavelengths, while the VHF band is named because of its frequency. It would have been more logical to call it the very-short-wavelength band. It illustrates that it doesn't matter whether we talk about wavelength or frequency, the two are always related by: wavelength times frequency equals speed of light.

Radio waves can be transmitted and received over the whole range 100 kHz to 100 000 MHz, the radio band, or wavelengths 3 millimetres to 3000 metres. The bands on your radio receiver are simply those that have been set aside for radio broadcasting. Other frequencies are set aside for television, communications, and radar. A few very narrow bands have been preserved for radio astronomy, centred at frequencies 38, 178, 408, 1400, 2700, and 5000 MHz. Nobody else is supposed to use these bands, since the faint signals reaching us from the cosmic landscape would easily be drowned by one small transmitter operating near the telescope. Astronomers have to battle continually to preserve silence in their bands. The worst offenders are the military authorities. Sometimes astronomers have been able to persuade the military to move out of their bands by threatening to publish the locations of the transmitters. There's even a nasty moment, for astronomers at least, in the film *You Only Live Twice* when James Bond takes off in a helicopter with the words 'You can reach me on 410 megacycles'.

Your radio receiver, though not as versatile a detector of light as your eye, is still quite elaborate. There are really two or three separate receivers in the same box, one for each waveband. When you push the button selecting the waveband you are really switching to a different receiver. Within any band you can also tune to any particular wavelength (or frequency) in the band. In the visible band of wavelengths, those visible to the eye, the different wavelengths or frequencies are the colours. The rainbow, for example, is a display of the different wavelengths in the visible band, spread out like the wavelengths on the tuning dial of a radio receiver. At the red side of the rainbow are the longer wavelengths (lower frequencies) and at the blue side are the shorter wavelengths (higher frequencies). The eye responds to wavelengths between 0.4 and 0.7 micrometres (1 micrometre = 1 thousandth of a millimetre). This is quite a narrow relative range of wavelengths, comparable to the relative range of one of the radio-broadcasting bands. The boundaries between the colours of the rainbow are not sharp. We talk of the seven colours of the rainbow — red, orange, yellow, green, blue, indigo, violet — but those who need to (for example painters) can learn to recognize hundreds of shades of colour. It is a beautiful experience to look at the artificial rainbow created by the prism spectrometer and let the whole range of colours slowly pass the eye.

Why has nature evolved this detector which responds to just this narrow range of frequencies? The answer is simply that this is the range at which the sun is at its most brilliant. By comparison the amount of sunlight emitted in other bands is almost negligible. However there might have been some advantage for nocturnal creatures in being able to see in the infrared, since this is the range of wavelengths in which the earth itself, plants, and other animals, radiate. Yet it seems that only a few snakes like the rattlesnake have managed to evolve a pair of infrared eyes. This is curious since the visible eye is said to have evolved independently in several different species.

When we close our eyes and switch on the radio, we are switching from one waveband to another. We can't ourselves detect radio waves so we translate them to a form which we can understand — in this case sound. Perhaps this is an odd way of looking at radio. Normally we think of radio waves as simply a device for carrying the sound to us over long distances, just as the ultra-high-frequency radio waves used to transmit television are a device for bringing pictures to us. But if we want to get an idea of what the world would look like if we had radio eyes instead of visible-light eyes, we have to think of a radio receiver as a device for translating the invisible radio waves into comprehensible form.

Now let's ignore the fact that structure has been imposed on broadcast radio waves so that the signal can be interpreted as music or the human voice (the waves are 'modulated' either by varying the amplitude of the waves — AM — or the frequency — FM) and just think of broadcasting as signals telling us where major transmitters are. Assume we have one of those portable VHF radios which are very sensitive to the way you point the set. We turn the tuning dial to some frequency and rotate the set until we get the loudest signal. This gives us the direction of the strongest transmitter at that frequency (it could be a small, near-by one or a powerful distant one). We then change frequency and try again. If we moved to a different spot and repeated the test, we could acutally pin down the distance of the transmitter by the surveyor's method of triangulation. In this way we build up a picture of what our locality looks like to a radio eye, the radio landscape. Every few hundred miles we find a very powerful transmitter — a broadcasting station. Then we might find a more localized but still quite strong transmitter with, at the same frequency, hundreds of much weaker transmitters moving around at about 20 miles an hour in zigzag paths — a radio taxi service. Finally we might also find another localized transmitter communicating with thousands of very weak transmitters moving about somewhat irregularly at a few miles an hour — police HQ talking to their men and women on the beat.

The whole radio band is a babble of voices saturating the earth's environment. Now suppose we turn to one of those frequencies set aside for radio astronomy. Suddenly there is total silence. Only if the aerial is enlarged to the size of a radio-telescope and very special receivers are installed do we start to see the radio cosmic landscape.

What is colour?

Colour means the same thing as the wavelength or frequency of light — provided we are talking about a pure colour, i.e. the light of a single wavelength or frequency, as in the rainbow. Where the light is a mixture of different wavelengths, we have an impure colour. The human eye does not work in terms of pure colours, and cannot distinguish exactly whether a colour is pure or not. However the brain analyses the information from the eye to determine a scale of colour purity, the tint. As a colour becomes more and more impure, and more light of other wavelengths is added, the colour becomes whiter and whiter. When we mix together similar amounts of light of all visible wavelengths we see a perfect white.

The light with the purest colour is that emitted by an atom when an electron orbiting round the central nucleus of the atom jumps from one orbit to another. The atom is the basic constituent of matter and consists of a tiny nucleus, positively charged electrically, with a cloud of negatively charged electrons orbiting around it. The nucleus can be broken down, under the extreme conditions of a nuclear accelerator, into positively charged protons, neutrons (with no electric charge), and many other strange particles with the function of holding the nucleus together. One of the main discoveries of quantum theory is that the electrons are only allowed to occupy orbits at a series of distinct distances from the nucleus. The further out the orbit, the more energy the electron needs to stay there and vice versa. An electron has a certain chance of jumping down to a less energetic orbit spontaneously, or it may get knocked down if the atom collides with another one. When the electron changes orbit from a higher energy orbit to a lower energy one, it has some excess energy to dispose of and it does this by emitting a particle of light, a photon. One of Albert Einstein's great discoveries, and the one for which he received the 1921 Nobel prize for physics, was that as well as being a wave-motion, light also has to be thought of as a stream of particles or photons. These photons have no mass but they do carry a certain amount of energy. The frequency of the light is proportional to the amount of energy the photon is carrying. From a particular type of atom, we may therefore see light at a series of characteristic frequencies corresponding to the various possible transitions from one permitted electron orbit to another. The atom can also absorb light with these same characteristic frequencies, in which case an electron jumps up to a higher energy orbit.

Practical examples of devices which make almost pure colours in this way are mercury or sodium street lights. They consist of glass bulbs with these vapours in them, heated up by an electrical filament. The mercury vapour radiates at a characteristic green wavelength and the sodium vapour at a characteristic orange wavelength. However these colours are still not perfectly pure. The individual atoms in the gas are moving around with respect to each other and with respect to our eye and this produces small shifts in frequency and wavelength, due to a very important effect pointed out in 1842 by the Austrian mathematician, Christian Doppler. When a whistling train passes us at a crossing, the pitch of the whistle goes through a sudden change from a higher pitch to a lower pitch. As the whistle is approaching us the sound waves coming to our ear get bunched together, because the train catches up a bit on the wave it has just emitted before emitting the next one. The wavelength is shortened, so the frequency or pitch is raised. As the train recedes, the waves are spread out, wavelength is lengthened and pitch or frequency drops. The same effect can be noticed in the pitch of a fast moving car engine as it passes us at the roadside. 'Frequency times wavelength equals speed of wave' works for sound as well as light, of course, as it does for all wave motions. The speed of sound is a good deal slower than light though, about 300 metres per second instead of 300 000 kilometres per second. The resulting relative shift in pitch (or frequency, or wavelength) is just given by the ratio of the speed of the source of sound (or light) to the speed of sound (or light).

A train moving away from us at 30 metres per second, 10 per cent of the speed of sound, causes a 10-per-cent downwards shift in pitch. The train would also change colour slightly, by 0.000 01 per cent in frequency or wavelength, since 30 metres per second represents this fraction of the speed of light. The eye doesn't notice this, of course, but shifts in colour almost as small as this can be detected on the surface of the sun using an instrument called a spectrograph to study the thousands of pure colours emitted by the different atoms in the surface layers of the sun.

In a mercury or sodium gas lamp, the light comes from many different atoms each moving with a slightly different speed with respect to our eyes. The result is that instead of seeing the very precise green or orange frequency of the mercury or sodium atom, the frequencies are smeared out around the expected value. The amount the frequencies are smeared out depends on how fast the atoms in the gas are moving around. This in turn is determined by the temperature of the gas — the hotter the gas the faster the atoms whizz around. The temperature is just a measure of how much energy the atoms have on average, as they move around the gas.

I mentioned the rainbow as a good natural example of the pure colour scale. Another example of a natural colour scale is the colours of a heated furnace. As the temperature is increased the furnace glows red, then orange, then yellow. We are then near the melting temperature of most metals, but if we look instead at a hot cloud of gas like a star and keep on increasing the temperature then we would see first green, then blue as the temperature is increased. These are not pure colours however and they are so impure that even the eye can tell this. As we get towards the hotter end of the scale the colour looks whiter and whiter. A gas cloud hotter than about 40 000 °C looks completely white to the eye.

These colours are very important in astronomy, because when we see them we know that we are looking at hot gas and we can tell the temperature from the colour. A hot gas, or a perfectly absorbing surface like the inside of a heated furnace, emits light with a very characteristic distribution of wavelengths. The peak frequency, at which the emission is brightest, is proportional to the temperature[1] of the emitting body. So if the temperature is doubled, say from 4000 to 8000 °C, the peak frequency doubles too, from the red to the blue part of the band. The relative spread of wavelengths about the peak wavelength, which makes the colours look impure to the eye, is the same for all temperatures. The brightness of the light drops off very rapidly towards higher frequencies and less rapidly towards lower frequencies. Because this type of radiation is typical of a black, perfectly absorbing surface, it is often called 'black-body' radiation. However it does seem rather confusing to call a yellow object like the sun a 'black body', so I shall usually speak of 'thermal' radiation.

Before I explain how the eye sees colour, let me give the astronomer's definition of colour. The astronomer needs a definition that is independent of the human eye and will work at the invisible wavelengths too. We choose any two wavelengths (or frequencies or pure colours) say 0.65 micrometres in the red

---

[1] For this proportionality to be exact, the temperature has to be measured from absolute zero (–273 °C).

and 0.4 micrometres in the blue. We then measure the brightness of the source of radiation we are interested in at these two wavelengths and calculate the ratio of these two brightnesses, say red brightness divided by blue brightness. Pure blue would give a ratio zero, pure red would give infinity, and objects in the real world give numbers in between. The astronomer now has an objective number which describes the relative redness or blueness of things. To completely describe the light from the source we need to measure the 'colours' for a wide range of pairs of wavelengths, in theory an infinite number. For many purposes three or four bands are sufficient, the most commonly used bands in the visible range being in the red, green, and blue. To get a reasonable amount of light through onto the light detector or photographic plate we allow through quite a spread of wavelengths centred on the chosen wavelength, using coloured filters. It is very like the principle of colour photography where pictures of the world are taken through three coloured filters and then recombined to reconstruct the original scene.

And this is also the principle on which the colour vision of the human eye works. There are three types of cone-shaped detector on the back of the retina, containing organic pigments sensitive to red, green, or blue light. The pictures seen by these three sets of colour-sensitive detectors are combined by the brain to give the experience of colour. It is amazing how vivid colours seem given that they are determined from these three brightnesses. We can say that we see three dimensions of colour: one dimension could be taken as the average colour or hue, the second as the tint (a measure of how impure the colour is), and the third as the intensity of colour (for light which was a mixture of all wavelengths, this would be the scale of the whiteness or greyness of things). The latter is not a true dimension of colour, in the sense of the astronomer's definition of colour. In reality there are an infinite number of dimensions of colour, corresponding to the infinite number of possible pure colours, so the eye, like the colour photograph, perceives colour in a very approximate way. By contrast the ear can recognize over a thousand pure sounds or pitches. The brilliance of the human visual system lies in the ingenious programming by which the signals from the simple detectors are decoded by the brain.

In our voyages through the cosmic landscape we have to use a more general concept of colour than that of the eye alone since we shall be travelling mainly through the invisible wavelengths. So we shall use the astronomer's definition of colour given above, testing whether a source is stronger in one wavelength than another. One of the main things we will want to decide from colour is whether the radiation we are seeing is from a hot body — 'thermal' radiation — or whether some other, 'non-thermal' process is involved. One of the most common examples of the latter is the movement of particles close to the speed of light.

The all-frequency light machine

Let's imagine that we have a receiver that can tune to all possible wavebands — the all-frequency light machine. We transport it outside the earth's atmosphere and away from man-made interference, say to the moon. Each band is connected up to a suitable telescope which can instantaneously scan the whole sky. The results are fed to a special pair of spectacles so that you can see the sky as it appears to the telescope working in any particular band. What do you see as you tune across the wavebands? We shall leave explanations of the phenomena and the origins of these different radiations till later chapters. For the moment we concentrate on the appearance of the sky only.

We start with the visible band — that familiar sky of stars. We notice one thing immediately: the stars are not twinkling since there is no air to refract the starlight. We also see that the stars have different colours: most of the brighter ones are blue, but there is a good sprinkling of green, yellow, and red ones. The colours of the stars, which correspond to their different surface temperatures, can be seen with the naked eye from a good site on earth, for example a dry mountain site like Kitt Peak in Arizona. With a big telescope the stars appear as small pools of coloured light and from above the atmosphere they will appear as bright points of light.

That ribbon of light around the sky, the Milky Way, becomes resolved by our telescope into a giant disc of stars, our Galaxy. Those two fuzzy clouds of light noticed by Magellan in the southern sky are focused into other smaller star systems, galaxies orbiting our own Galaxy. Other more distant galaxies swim into view, some isolated, some in small groups, and some in huge swarms or clusters.

Now we shift the waveband past the blue end of the spectrum, past the violet and into a band of invisible radiation, the ultra-violet. The red, yellow, and green stars start to fade out, including the sun, while the blue ones become even brighter. Many of the galaxies, composed mainly of red stars, start to fade too. But in some we start to notice something interesting happening at the very centre of the galaxy, the nucleus. An intense point-like source of light starts to become more prominent. These are the 'active' galaxies, so called because of the evidence for explosions and other violent activity in their nuclei. The most intense of all, those in which this point-like source dominates the output even of all the stars in the galaxy, are called 'quasars', short for quasi-stellar radio source. These were discovered as star-like objects at the position of bright radio sources and only gradually realized to be the most distant and luminous objects in the universe. As we switch our waveband into the extreme ultraviolet a fog descends. Quasars, galaxies, the Milky Way, all fade from view and only the very nearest of the blue stars can be seen. The gas spread between the stars of the Milky Way has mopped up most of the light before it reaches us. This dark night of the far ultraviolet continues until we have increased the frequency (or shortened the wavelength) by another factor of ten. We have reached the 'soft' X-ray band and the fog turns to a light mist. Through the mist we see a different world, the world of gas at a temperature of a million degrees or more. The blue stars have almost faded out. Instead some pairs of stars that we had hardly noticed before blaze out. One of the stars is close to death. The Milky Way has returned, but instead of a broad river it has become a narrow stream. Into the 'hard' (higher frequency) X-ray band and the mist lifts completely. The Milky Way disappears as we explore the world of gas ten or a hundred million degrees centigrade or hotter. The dying stars in double-star systems are still there. So are the quasars and active galaxies, but normal galaxies are very weak. And a new kind of source is seen: the rich clusters of galaxies are bathed in diffuse light. We try to switch our all-frequency light machine to higher frequencies, the gamma-ray band, but the picture becomes very blurred and crude. We have reached the high-frequency limit of present-day astronomy.

What happens if we return to the visible band and switch in the opposite direction, towards lower frequencies and longer wave-lengths. As we move from the red end of the visible band into the invisible 'near' (shorter wavelength) infrared the blue stars fade out and myriads of red stars swim into view, red giants of high mass and red dwarfs of low mass. Switching towards the middle infrared the stars start to fade out and cooler objects — planets, asteroids, and comets — become more noticeable. We are in the band dominated by dust warmed by stars — the interplanetary dust that causes the zodiacal light, cometary dust, and that precious speck of dust, earth itself. Certain of the red stars become prominent again because of the radiation from clouds of dust surrounding them, the cooled relics of material thrown off by the stars. In some cases this dust cocoon is so thick that the star hardly shines through at all, so we notice them for the first time in this band. We move to the 'far' (longer wavelength) infrared, to the world of cold matter (–200 °C or less). The stars have all vanished, even most of those shrouded in dust. Instead we see huge dimly glowing clouds of gas and dust spread round the Milky Way. In some, stars may just be beginning to form out of the gas. The nucleus of our own Galaxy, in the direction of the constellation of Sagittarius, the Archer, stands out because of the great concentration of stars there, buried in thick clouds of dust. The nuclei of some other galaxies are even more dramatic, in some cases because of a burst of newly formed stars, in others because an active nucleus is hidden within the dust.

Between the longest far-infrared and the shortest radio wavelengths, in the 'microwave' band, we become aware of a faint sea of light all round the sky, the 'cosmic microwave background' radiation, relic of the 'Big-Bang' origin of the universe over ten thousand million years ago. The gas clouds of the Milky Way begin to be illuminated in the vicinity of bright blue stars that have recently formed out of the gas of the clouds. We notice the quasars and active galaxies becoming bright again. And at certain characteristic wavelengths the world of interstellar molecules lights up.

Into the radio band and the Milky Way has spread out into a bloated avenue of light, much brighter towards the direction of the constellation of Sagittarius than in the opposite direction, the constellation of Taurus the Bull. The plane of the Milky Way is still picked out with radiation from hot gas near young blue stars. The rest of the sky is rather uniformly covered with fainter sources, mostly active galaxies and quasars. Other near-by normal galaxies seem very weak by comparison.

As we move to lower radio frequencies we notice a smooth

haze of background radiation coming from the whole sky. By comparison the Milky Way, which has spread out to cover almost the whole sky, starts to fade and the clouds of hot gas, which were so prominent at high radio frequencies, suddenly vanish and turn into dark patches. The same clouds of hot gas are now absorbing the radio waves from behind them. Thus high and low radio frequencies are like positive and negative copies of a photograph, with bright sources on a dark sky at high frequencies and dark holes on a bright sky at low frequencies. Many of the sources spread over the sky have faded too and the sky is becoming more sparsely populated than at any time since we passed the gamma-ray band, apart from the fog of the extreme ultraviolet. Before we can decide at what frequencies these sources too fade, our receiver has begun to crackle and hiss. We have again reached the limit of the known world.

But what exactly is this all-frequency light machine? It is none other than we ourselves, and these telescopes are our eyes. New eyes that are the product of our cultural evolution.

How is light made?

Consider the journey that sunlight has made before it falls on the landscape in front of me now. It originated in thermonuclear reactions deep in the sun's interior. Thermonuclear reactions, which are what power the H-bomb too, involve the transmutation of one element into another. Our alchemical transmutation starts with four of the simplest possible kind of atom, the hydrogen atom, with one electron orbiting round one proton. These four atoms fuse together (after various subtle interactions) to form the next most complicated atom, helium, with two electrons orbiting round a nucleus of two protons and two neutrons. Now a helium atom weighs a bit less than four hydrogen atoms and part of this difference appears — in an awesome demonstration of the principle discovered by Albert Einstein that matter and energy can be converted from one to the other — as energy, in the form of gamma-ray photons. These gamma-ray photons do not travel far before being absorbed by other atoms, changing the atoms to a higher energy state. These atoms soon release this extra energy by emitting new photons, each of lower energy, say in the X-ray band. This process occurs many times as the energy travels out towards the sun's surface. Finally a visible photon emerges from the sun and rushes unmolested through the thin gas of the interplanetary space between us and the sun.

The sun's light, however, is not unmolested by the earth's atmosphere. If it was we would see a bright sun and a black sky, like the astronauts do. Why do we see a bright blue sky? As the photon travels through earth's atmosphere there is a small probability that it will be absorbed by a molecule of air. More likely it will be randomly reflected or 'scattered' off a molecule of air or a small particle of dust suspended in the air. A small particle is most effective at scattering light of wavelength smaller than the size of the particle. As the wavelength gets longer than the size of the particle, the effectiveness of the particle as a scatterer declines rapidly. In fact a molecule of air or a small dust particle is about sixteen times more effective at scattering blue light than it is at scattering red light. Few blue and violet photons reach us without having been scattered several times. When they reach us they come from the direction of the molecule that scattered them last, so they can come from any direction on the sky and the sky looks blue in every direction. The blue sky is nothing more than the blue light of the sun rerouted to our eyes by dust and air molecules. This was first realized by Leonardo da Vinci in 1498, who noticed that the sky became much darker as he climbed Monte Rosa in the Alps and who also made the connection with the blue appearance of wood smoke seen against a dark background.

The red and yellow photons from the sun tend to come directly and unmolested. The removal of the blue photons means that the sun has been 'reddened' compared with the light reaching the top of the atmosphere. When the sun is low in the sky, so that sunlight has a much longer path to follow through the atmosphere, the reddening of the sun becomes very noticeable. We even start to see some scattering of yellow and red light into a sunset halo round the sun. After a volcanic eruption, which fills the atmosphere with dust particles for a few days, the sun looks red at noon too.

Finally, after their long journey from the sun, all but the green photons are absorbed by the green leaf of tree or plant and so only the green photons are reflected to my eye. When the photons reach my eye they are absorbed in the cone detectors of the retina. The brain easily determines that rather few of the photons are stimulating the red- and blue-sensitive cones and that the majority are stimulating the green-sensitive ones.

This story is grossly oversimplified. There are many different kinds of interaction between light and matter which come into play at different levels of the sun's interior. The surface layers are especially complex, since the atoms of almost every element in the periodic table manage to imprint their signature on the colour of the emergent light. Yet we are not so far from the truth when we say that the sun looks like a hot ball of gas at about 5600 °C. Most of the escaping photons carry about the same amount of energy as the atoms in the sun's surface layers. To carry this amount of energy, their frequency has to be that of yellow light and so we see our familiar yellow sun.

However the sun emits other types of light besides this basic 'thermal' radiation. Above the visible surface there exists a zone of thin gas at about one million degrees centigrade, the corona, and this emits 'free–free' radiation in the ultraviolet and X-ray bands. This is radiation that is emitted when a free electron (one that is not bound into an atom) is strongly deflected in the electric field of a proton or positively charged atomic nucleus. The gas must have electrons stripped off a fair proportion of its atoms by collisions between atoms for this to happen and for hydrogen this means it must be hotter than about 20 000 °C.

The sun also emits radio photons which are formed by a totally different 'non-thermal' mechanism. An electron accelerated to nearly the speed of light in a solar flare or disturbance, spirals in the sun's magnetic field giving off energy in what is called 'synchrotron' radiation, so called after the man-made particle accelerator where the radiation was first discovered. The main difference in appearance between a source radiating thermal and non-thermal radiation is that thermal radiation is strongly peaked to a narrow range of wavelengths, whereas non-thermal radiation tends to be spread over a wide range of wavelengths.

I shall say more later about these different ways that light is made and unmade. Recognizing these radiation mechanisms is one of the astronomer's major tools in understanding how the sources he or she sees work. We do not just want to contemplate the cosmic landscape, but also to understand how it was formed and to find there the clues to our own past.

How does the cosmic landscape involve us?

We have all looked up into the night sky and asked unanswerable questions: Does it stretch to infinity? Has it been there for all time? How did this earth, and we ourselves, come to be here? Are we alone in this immensity of space and time?

Astronomers have made efforts to answer all these questions, with only limited success. For example the universe seems to have originated in a dramatic explosion 20 thousand million years ago — the 'Big Bang' — but there is little progress in deciding whether it is of finite or infinite extent. There have been searches for signals from civilizations orbiting near-by stars, but the lack of success is not necessarily surprising even if advanced life is common in our Galaxy. There are several possible scenarios for the origin of the solar system and a few ideas about the origin of life, but no convincing and widely accepted picture.

Yet the cosmic landscape does involve us in a particularly direct way. Almost every atom in our bodies has been through the furnace of a star's interior. Apart from hydrogen and helium, the elements are formed in the stars, built up by nuclear reactions from simpler elements and originally from hydrogen itself. The stars that made our atoms died explosive deaths long since but we can see others like them in the cosmic landscape. We can see, too, examples of their dramatic deaths and of the tiny remnants they leave lit up in X-rays. The study of the evolution of the stars, from their birth from clouds of gas to their death in a shower of exploding gas, is therefore part of our own evolution. Between the stars we seen the clouds of gas forming and settling, ready for the cycle to begin again.

And spread through this interstellar gas we detect small grains of dust, one ten-thousandth of a millimetre across, composed like miniature rocks of silicates or graphite. These grains are the storehouse of the elements from which the earth was formed. In recent years an even more direct pointer towards our origin has been found. The denser clouds of gas and dust are rich in simple organic molecules, from carbon dioxide to ethyl alcohol, which are the first steps towards life's complex molecules. We are one with this cosmic landscape, for a human being is a bag of molecules moving around on the surface of a grain of dust.

On a larger scale we find that the stars we can see are part of a huge rotating disc, the Milky Way galaxy. With large telescopes we find other similar systems at great distances, often congregated in clusters of thousands of galaxies. These systems were formed early in the evolution of the universe and allow us to probe the origin of the cosmic landscape we are a part of.

We can search back to an even earlier epoch, before the galaxies were formed, by studying the microwave background radiation left over from the fireball phase of the Big Bang. With the slenderest of evidence to guide them, the theoreticians struggle to make sense of the earliest years and seconds of the universe's history.

How will earth itself seem to us when we return from our voyages through the cosmic landscape? Naturally it will seem more than ever a secure garden for mankind. For nowhere else do we yet find other life in the universe, nor even certain evidence that the necessary preconditions exist. And however much we can explain, the very fact of our existence remains miraculous. As Wittgenstein says: 'Not how the world is, but that it is, is the mystical experience.'

And the knowledge we have gained of this landscape itself makes earth the most signigicant cosmological object we know. Perhaps here alone the universe achieves some consciousness of itself. Aristotle's picture, with the earth at the centre of the cosmos, does not seem out of proportion.

Voyages back along the photon's track

The astronauts' first voyage to the moon is a memorable experience of our time. Yet the possibility of travel to even the nearest star remains remote. As for journeys across the Milky Way or to another galaxy, these are almost beyond imagination to us. Yet light from these distant places has made the journey to earth and it falls on these new eyes of ours, the telescopes, the all-frequency light machine.

And with these new eyes we can make voyages as dramatic as those of the early navigators, out to the limits of the known universe. As we look out to ever more distant parts of the cosmic landscape, the light that is reaching us now set off to us ever further back in the past. As we have seen, the light from the nearest star set off four years ago and from the most distant sources known to us over ten thousand million years ago, long before sun or earth were formed. When, with our new eyes and our imagination, we embark on our voyage across these dark spaces, we start to travel back through time. Like the archaeologist's strata, the cosmic landscape contains examples of every stage of our past. Interpretation is difficult, and harder the further back in time we travel. As we draw together the lines of evidence from these eyes of ours working in their different wavebands, a picture of the history and evolution of matter in the universe grows. Of course, being only human there is a tendency for us to create a picture out of the slenderest evidence of our senses and then to believe totally in the picture, and to forget how slender is the evidence. Cosmologists talk learnedly about the first fraction of a second of the universe's history, when in fact we have no really direct evidence relating to earlier than a million years after the universe's apparent birth in the Big Bang.

In this book we follow a different approach. We leave on six voyages through space and time, corresponding to the six main wavebands of the all-frequency light machine, back along the photon's track. We explore the cosmic landscape as it appears to eyes working in these wave bands. And although I shall say something of the explanations that astronomers have concocted for these phenomena, I shall leave most such explaining to your own imagination, and concentrate on showing you the landscape as it appears to me.

The rivers of light flow to our eyes, and we shall trace them back to their source.

2
FIRST VOYAGE
The visible landscape

We are ready to embark on the first of our voyages through the cosmic landscape, back along the tracks of photons. The first steps of this voyage are along a familiar road for we are to travel in visible light. In the foreground we see the trees, rocks, and rivers of our home planet. How comfortable and solid they seem as we gaze at them for the last time before departing. Will they seem the same to us when we return?

The sun

Whichever of the photons entering our eye from this terrestrial landscape we choose to follow back to its source, we are led in one direction only: to the sun, source and sustenance of life on earth. When we leave the top of earth's atmosphere after numerous scatterings, we are heading towards the sun, 8 light-minutes away. As we approach the blazing surface we see that it is far from being the smooth surface it appears from earth. It resembles a boiling cauldron, a seething honeycomb of rising and falling cells of fluid. This 'granulation' of the sun's surface shows that near the surface of the sun, energy is transported by convection, just as it is through the water of a boiling saucepan. Sometimes streamers of hot gas, called 'prominences', leap up from the surface like a fountain or a volcano in eruption before falling back in a spectacular arch. Small dark spots grow on the surface, lasting for several days. These mark regions where the magnetic field of the sun's surface layers has become tangled up. We see that the sun is slowly rotating, once every thirty days. Occasionally we may see part of the sun light up in a dramatic flare, again associated with magnetic disturbance.

The first signs of the sun's activity to be discovered were the sunspots. They can easily be seen with the naked eye by projecting the image of the sun onto a white surface. One of Aristotle's pupils, Theophrastus of Athens, is said to be the first person to notice them, in 300 BC. The Chinese astronomers carefully

23

recorded their occurrence between the first and seventeenth centuries AD. Sunspots were first observed with a telescope in 1611 by several people, including Galileo. In 1843 the German astronomer Heinrich Schwabe found from his long series of sunspot observations that the number of spots on the sun waxes and wanes over an eleven-year period. Other types of activity (prominences, flares) also vary in step with the sunspot cycle. Even the annual growth of trees on earth changes in phase with the cycle — it can be seen as variations in the width of tree rings. It was realized early in the present century that between 1645 and 1715 the sun behaved strangely, or rather, it stopped behaving strangely. Between these dates, which coincide with the 'Little Ice Age' in Europe, there were almost no sunspots at all. Aurorae and magnetic storms, of which more later, also stopped. Then in 1715 the eleven-year cycle started up and has continued to the present. This 'Maunder' minimum, named after the British astronomer E. W. Maunder who drew attention to it in 1904, and its connection to the earth's climate, remain unexplained.

The inner planets

We leave the sun and search out its tiny companions, the planets. The first two, Mercury and Venus, we can occasionally see in twilight. For the remainder, and for the rest of the voyage, we have to travel by night. Daylight is the enemy of the astronomer who works at visible wavelengths.

Mercury resembles nothing so much as a battered stone, bruised and scarred by the impact of giant meteorites over the thousands of millions of years of the planet's life. The Mercurian day lasts 59 earth-days, exactly two-thirds of its year of 88 earth-days. This ratio is not a coincidence, but is an effect of the sun's gravity on Mercury. The surface of Mercury is a harsh environment. The sun looks three times larger than it does on earth and is ten times brighter. At noon the surface temperature of Mercury reaches over 400 °C. At midnight (29½ days later) it drops to −170 °C.

The surface of Mercury is, in fact, not saturated with large craters in the way that the moon is. There are conspicuous plains which pre-date most of the large impact craters. Another feature unique to Mercury is the ubiquitous presence of shallowly scalloped cliffs and scarps running for hundreds of kilometres, probably due to the contraction of the crust as the planet's large iron core formed.

The next planet, Venus, is a strange world of red-hot rocks and mountains with a massive atmosphere of carbon dioxide and a layer of thick clouds. In the visible band we see only the feature-less yellow tops of these clouds, at a temperature of –20 °C. Like the moon and Mercury, Venus shows prominent phases due to the sun shining only on the half of the planet nearest to it. The Venusian day is 243 earth-days long and its rotation is in the opposite sense to the earth's, so the sun rises in the west and sets in the east. Whenever Venus is 'new', i.e. between the earth and the sun (this is called the 'inferior conjunction'), it always presents the same face to earth. This suggests that earth's tides there have 'captured' Venus's rotation, though in a less dramatic way than the moon's, which is forced always to show the same face to the earth.

The voyages of the astronauts have made the landscape of the dusty, cratered moon and its distant blue planet swathed in white clouds, earth, familiar to us. Why does earth escape this fate of Mercury and the moon to be covered with craters? The reason is that the main period of crater formation by meteorites occurred thousands of millions of years ago, soon after the solar system was formed. Erosion by wind and rain, and the movements of the earth's crust, have long ago obliterated these early craters on earth. Mercury and the moon, on the other hand, have no atmos-phere and none of the violent crustal activity of earth. The inert moon carries the scars of all its meteorites and bears witness to an era of gigantic meteorites with its immense *maria*, which Galileo thought were seas. Even the minutest grains of interplanetary dust leave their tiny marks on the airless moon, and some of the lunar rocks brought back by the astronauts are pitted with miniature craters visible only under a microscope.

However the eroded remains of giant meteoritic craters can be found on earth too, for example the ring-shaped Lake Manicougan in Canada. This crater is 70 kilometres in diameter and was caused by an object three kilometres across, an asteroid, hitting the earth about 200 million years ago. Such colossal impacts are rare. The last really dramatic meteorite was seen by a few amazed observers in Siberia in 1908. The fireball of the explosion flattened trees up to 30 kilometres away. About 1500 reindeer were killed and a man standing on the porch of his home 70 kilometres away was knocked unconscious. In 1972 a 1000-ton meteor skimmed off the earth's atmosphere above Montana, in the USA, like a stone skimming on water, and out into space again. (Out in space, rocks, stones, and grains of dust are called meteoroids. When they blaze out in visible light as they travel through the earth's atmosphere and heat up because of friction, they are called meteors. And if they survive to hit the ground they are called meteorites.) Several fist-sized stones, and tons of dust grains, hit the earth's surface as meteorites each year. Usually when one of the smaller grains runs into the earth's atmosphere, the friction of the grain's race through the air heats it up so much that for a moment it blazes out in the visible band before it melts away. It becomes a shooting star, a meteor. At certain points on its orbit the earth runs into a denser cloud of dust and we can see one of the famous meteor showers, the Leonids or the Perseids. All night the sky is crossed by shooting stars.

When one of the larger rocks enters the earth's atmosphere a more dramatic display takes place. A ball of fire plunges across the sky as the rock is heated up, followed by a loud explosion as it crashes into the ground. A meteorite has landed. It is amusing to recall that it took astronomers a long time to believe the popular reports of stones falling out of the sky and only in 1803, when a meteorite exploded over the French town of L'Aigle was the issue settled conclusively. Large meteorite falls are not common and usually happen over the sea or unpopulated areas. When the rocks are seen to fall and are recovered, scientists naturally treasure these examples of extraterrestrial material. Stones which 'fell from the sky', almost certainly meteorites, were venerated at Delphi and Ephesus in classical times, at Mecca, in ancient Mexico, and by prehistoric Indians of North America.

Two famous meteorites are the Orgeuil meteorite, which fell in France in 1864, and the Murchison meteorite, which fell near Murchison, Australia, on 28 September 1969, and portions of which were collected that day. Careful analysis of radioactive elements in another meteorite which fell in 1969 near Allende, Mexico, has shown that some of the material in it was 'cooked' in a star only shortly before the solar system formed.

Meteorites can be divided into two main classes: stony and iron meteorites. The stony meteorites, or chondrites, have a particularly interesting sub-group called the carbonaceous chondrites, of which the Orgeuil and Murchison meteorites are examples. When analysed in 1961, almost a century after its fall to earth, the Orgeuil meteorite showed traces of organic material. Could this interplanetary jetsam carry alien life to earth? After so long a delay it was hard to be sure that this organic material was not due to terrestrial contamination. A far more convincing demonstration of organic material in meteorites followed the fall of the Murchison meteorite, which was found to contain eighteen different amino acids, the building blocks out of which living cells manufacture protein. These could not be terrestrial contaminants since half of the material in the form of any one particular amino acid had a molecular structure which was a mirror image of the form found on earth. These organic molecules in the Murchison and other stony meteorites are believed to be formed in the kind of chemical synthesis which must have occurred on the young earth. Did meteorites like the Murchison ever carry living organisms? Most astronomers doubt this, but recently Fred Hoyle and Chandra Wickramasinghe have argued (in *Life cloud*) not only that they did, but also that these organisms could survive the heating of the passage through earth's atmosphere. They connect world-wide epidemics with the earth's passage through clouds of meteoritic dust. I shall return to the question of the origin of life in the last two chapters.

After earth and its cratered moon comes Mars, a world of deserts of red sand, of polar caps of carbon-dioxide ice, of thin atmosphere, of winds and sandstorms, of eroded craters and dried up rivers. Mars is the nearest the solar system offers to a twin for earth, and conditions may once have been favourable for life. But there are no signs of life, past or present. What pleasure even the humblest moss or lichen would have given us.

The day on Mars is only 41 minutes longer than the day on earth, while its year is of 687 earth-days. The thin atmosphere is composed mainly of carbon dioxide, with traces of carbon monoxide, oxygen, ozone, and hydrogen. There is also enough water in the atmosphere to cover the surface of the planet with a liquid film only a hundredth of a millimetre thick. Frozen into the polar caps, however, there is enough water to cover the planet with water 10 metres deep, and there may be still more in the form of a permafrost layer under much of the planet's surface.

Mars has the largest volcano in the solar system, Olympus Mons, 600 kilometres in diameter and rising 25 kilometres above the surrounding plane. Erosion along the perimeter of the long extinct volcano has created cliffs 2 kilometres high. It must be a formidable sight. Just south of the Martian equator is a region of enormous canyons, dwarfing the Grand Canyon of Arizona, USA. The volcanoes and canyons of Mars are especially impressive when you think that the red planet is only a tenth of the mass of earth. Near the equator we see also the sinuous channels of many dried up rivers, which flowed long in the past when Mars was wetter and warmer. Two cratered and irregular boulders, 10 to 20 kilometres across, circle the red planet. They are the Martian moons Phobos and Deimos, possible fragments of a larger moon shattered by collision with an asteroid.

The asteroids

Beyond Mars telescopes reveal a belt of hundreds of thousands of rocks and boulders, many a kilometre or more across — the asteroids. As we pass them we sense some past calamity, the disruption of a tenth planet or, more probably, its failure to form. The largest asteroid, Ceres, was discovered in 1801 by the Sicilian monk Guiseppe Piazzi and is 955 kilometres across, almost a third as big as the moon. The orbits of about two thousand asteroids are known accurately and thousands more have less well determined orbits. The most interesting group is the Apollo asteroids, named after the first of them to be discovered in 1932. These are asteroids whose orbits take them inside the orbit of the earth, so that there is a possibility of them hitting the earth at some time. Apollo itself was lost from 1932 until its rediscovery in 1973, during which time it had crossed the earth's orbit many times. In 1937 the Apollo-type asteroid Hermes, one kilometre across, passed within 800 000 kilometres of the earth, less than twice the distance of the moon. It is expected that an Apollo asteroid will pass closer than the moon every century and that one will collide with earth every 250 000 years. Twenty-eight Apollo asteroids are now known, ranging in size from 400 metres to 8 kilometres, and there may be as many as a thousand in all.

It is the Apollo asteroids that are responsible for the giant craters, five kilometres or more in diameter, on the inner planets and the moon. Fragments of them resulting from collisions with other asteroids are probably responsible for most of the meteoritic debris raining on earth. From studies of ancient craters, we can estimate that about 1500 Apollo-type asteroids have hit the earth since the beginning of the Cambrian geological era 600 million years ago. The impact of an object like Hermes on the earth would be similar to the explosion of a 10-megaton H-bomb.

What is the origin of these fascinating but alarming objects? Some may have been deflected from the main asteroid belt by Jupiter's gravity. However most are probably the remains of comets, which we will be encountering shortly.

The outer planets

Now we leave the planets of rock and travel on to the planets of liquid, gas, and ice. The first, Jupiter, is easily the largest of the sun's retinue, 10 times larger and 300 times heavier than earth. In fact it comes quite close to being a tiny star, the defining characteristic of a star being that thermonuclear reactions take place in it. The centre of Jupiter, at 30 000 °C, is on the way to being hot enough to start nuclear reactions — if Jupiter were perhaps ten times as massive, the solar system might be a two-star system.

We see that Jupiter rotates about once every 10 hours and that the surface is divided into reddish or greenish brown bands separated by yellowish zones. These zones may result from phenomena similar to some of the meteorological phenomena in the earth's rotating atmosphere. The bands are broken up into smaller swirls of colour — oranges, reds, greens, and browns — and slowly change, like terrestial cloud formations. Slowly changing too is the prominent landmark of the Red Spot, discovered by the Italian astronomer Giovanni Cassini in 1665. It covers an area as large as the earth and is now believed to be a vast storm, which has already been raging for at least 300 years.

The interior of Jupiter is mainly liquid hydrogen and helium, possibly with a small molten iron core. In the surface layers we see crystals of ice and ammonia floating in hydrogen gas.

Looking down on Jupiter is like looking down on a miniature solar system, for there are fourteen moons. The four largest are the Medician moons, Io, Europa, Ganymede, and Callisto, 3000–5000 kilometres in diameter and discovered by Galileo during his first months with the telescope in 1610. Io has red polar caps of sulphur, a surface covered with active volcanoes and reddish brown evaporated salts (said to resemble a pizza), and at an atmosphere often aglow with a yellow aurora because of sunlight scattered off sodium atoms. Overhead Jupiter measures 18 degrees across — the size of the Plough as it appears in our sky. The other three Medician moons are covered with gravelly soil and frost.

On to Saturn with its beautiful rings of stone, dust, and ice, and its ten satellites. The largest of the satellites, Titan, has an atmosphere of hydrogen and ammonia and is almost as large as Mars. Saturn is the last of the planets that is visible to the naked eye. We travel on to the cold, greenish planets — Uranus, found in 1781 by William Herschel during his survey of the sky by telescope, and Neptune, discovered in 1845 by John Adams and a year later by Urbain Leverrier from its perturbations on Uranus's orbit.

The US astronomer Percival Lowell thought anomalies in Neptune's orbit showed the existence of a ninth planet six times as heavy as earth, but we now know his calculations were incorrect. Yet in 1930, Clyde Tombaugh, continuing Lowell's search for this hypothetical planet, did find a ninth planet — Pluto. The discovery in 1978 of a moon orbiting round Pluto showed that Pluto is little more than a large asteroid. It has a highly elliptical orbit and for twenty years of its 248-year orbit, from 1978 to 1998, it will lie inside that of Neptune. At the distance of Pluto, the sun looks decidedly pallid, only ten times brighter than the full moon looks to us on earth.

Beyond Pluto we come to the zone of the comets. Thousands of millions of these aggregates of rock, dust, ice, and frozen gases circle the solar system endlessly. Occasionally the gravitational attraction of a near-by star or a planet deflects one onto an orbit plunging in towards the sun. Only as it approaches the zone of the earth's orbit does it start to shine out in the visible band. The frozen gases vaporize, light up, and escape from the head of the comet into a fuzzy 'coma' around it. As the comet approaches the sun the pressure of the sunlight pushes some of the dust away from the head of the comet to give the spectacle of the 'tail' stretching out millions of miles behind it. Some, like Halley's comet, which returns every 76 years, are visible even in the day time. Others like Encke's comet complete their elongated orbit much more quickly (in a few years) and gradually fade as their gas and dust are driven out. The head of such a comet will end up looking like an Apollo asteroid and this is probably how most of the Apollo asteroids originate.

So far our voyage back along the track of the photons has taken us back in time only an instant, minutes for the inner planets or hours for the outer ones. But to reach the world of the stars we must travel back for several years. Our travels in the solar system are only a tiny step out into the cosmic landscape.

The zodiac and the constellations

As we take a last look back towards the solar system we see that the planets seem to be strung out on a line on either side of the sun, for they all lie roughly in one plane, the ecliptic or zodiacal plane. From the earth we see the planets trace out over many years a band round the sky, a great circle which marks where the ecliptic plane intersects the sphere of stars. And scattered about this band we find the bright stars which delineate the constellations of the zodiac.

The origin of the modern constellations is obscure. Many were first invented by the Babylonians or Greeks, some by the Romans, and others in more recent times to make sure that every part of the sky falls in some constellation. Different cultures have divided the bright stars up into constellations completely different to those we are familiar with. The ancient Chinese had 283 constellations, their average size being only a third of ours. By contrast, the Pueblo American Indians recognize a constellation which lies across almost the whole sky, in the form of a giant, the Chief of the Night.

The stars of a constellation are not really near each other in space, they just happen to be in nearly the same direction in the sky. The stars of the Plough or Big Dipper only form that pattern when observed near the sun. If we travel to a different place in the Milky Way, they would look completely different. Some would look brighter, because they are nearer, others would look fainter, and they would not necessarily be near each other in the sky.

The constellations are not much used in modern astronomy, except to give names to the brightest stars (they are labelled in approximate order of brightness, α, β, γ, and so on, then letter codes, and finally numbers are used), but they are useful to the ordinary star-gazer trying to find his way around the night sky. When I say something is 'in' a particular constellation, I mean that it is in that direction on the sky. It could be much further than any of the stars that make up the constellation, even beyond our whole Milky Way system. Because the ecliptic or zodiacal plane is so special, being the apparent path of the sun through the stars during the year, it has been carefully divided into twelve constellations, one for each month; the twelve constellations or 'signs' of the zodiac. How people imagine the direction of the sun on the sky at the moment of their birth is supposed to influence their personality and daily lives is beyond me.

As we look back towards the solar system from a distance we see that the sun and planets are enveloped in a faint disc of light, the zodiacal light. This can sometimes be seen from earth just after sunset or before sunrise as a faint, cone-shaped patch of light pointing up from the sun along the zodiac. Three hundred years ago Cassini suggested that this was due to sunlight reflected (or 'scattered') from particles of dust spread through the ecliptic plane, and this is the explanation believed today.

Giants and dwarfs

When we stand on earth and look up at the constellations, our eyes should be drawn to two of the brightest stars. In the southern hemisphere, the brightest star in the constellation of the Centaur, α-Centauri, is the nearest star to the sun. It is the third brightest star in the whole sky. In the northern hemisphere, the brightest star in the sky, Sirius, the Dog Star, is the fifth nearest star to the sun, only 8 light-years away. It lies in the constellation of Canis Major, the Big Dog. However the only other prominent stars visible to the naked eye to figure among the hundred nearest stars (those within 20 light-years) are Procyon, the brightest star in the constellation of Canis Minor, the Little Dog, and Altair, the brightest in the constellation of Aquila, the Eagle. Some of the other very near-by stars are faintly visible to the naked eye, but most can only be seen with a telescope. The stars we see with the naked eye are for the most part not the nearest but the most luminous in our part of the Milky Way galaxy. Most of the naked-eye stars are within a thousand light-years of earth, but they represent only a small fraction of the stars within that distance.

As we approach α-Centauri we see that it consists not of one star but of three stars, two of them yellow and rather similar to the sun, the third, a small, very faint red star. Stars significantly smaller than the sun are called dwarf stars, while those significantly larger are called giants. In addition, stars are characterized by their colour — red, blue, or white — depending on their surface temperature. Red means fairly cool as stars go, a few thousand degrees centigrade. Blue means about ten to thirty thousand degrees, and white means even hotter than that. Red dwarf stars, like this one in Centaurus only a twentieth the size of the sun and a ten-thousandth of its light output, or luminosity, are very common in the Milky Way, and most of the near-by stars are of this type. The weakness of their output explains why they do not appear among the prominent naked-eye stars. About half of the near-by 'stars' turn out to be the double-star or 'binary' systems and about a quarter have three or more members. α-Centauri is one of the best candidates among our neighbours for having earth-like planets. Any optical astronomers on an α-Centauran planet will not get much opportunity for work. Usually one or other of their two brighter suns will be up and sometimes both. Really dark nights when none of their three suns are up must be quite rare.

Sirius at first seems to be a single hot blue star twenty times more powerful than the sun, but in close up we see its faint companion, a white dwarf star known as Sirius B. Both these stars are much younger than the sun, yet the white dwarf is already dying, having exhausted its nuclear fuel. The blue star will not last more than a thousand million years in all, compared with the five thousand million years left to the sun (plus the five thousand million years it has already existed). It is more massive than the sun but it is being much more profligate with its resources of hydrogen gas.

Already we have spanned most of the range of star types, from luminous massive blue stars, through yellow stars like the sun, to faint low-mass red stars. Their colours show them to be hot balls of gas, from 2000 °C for the coolest and reddest to 50 000 °C for the hottest and bluest. These temperatures refer to the outside only: deep inside where the nuclear fusion takes place the temperature must be tens of millions of degrees. To find the other main type of star we go towards the constellation of Orion, the Hunter, and the bright red star Betelgeuse in the Hunter's shoulder. Betelgeuse is a red giant star, five hundred times larger than the sun and over ten thousand times more luminous.

We can now describe in broad terms the life history of a star after its formation. It spends most of its life fusing hydrogen into helium and while it is in this phase its mass determines the colour, size, and luminosity it adopts. It would be a red dwarf, a yellow solar-type star, or a luminous blue star, depending

33

whether its mass were, say, a tenth of, equal to, or ten times the sun's. But when the hydrogen near the hot centre is all consumed, the star goes through a rapid and spectacular phase of evolution. The star expands and its surface cools until it is red and bloated — a red giant. Meanwhile the core of the star has condensed to still higher density and temperature so that it can ignite a new fuel, helium, which is fused to form carbon, nitrogen, and oxygen. When the helium is exhausted, the core condenses further until the next fuel can ignite, and so on through the heavier elements up to iron. The star then explodes in a more or less dramatic manner and dies to form either a white dwarf like Sirius B or one of two other types of relic star which we will meet later, a neutron star or a black hole.

Stars that change and stars that explode

The stars are not all steady in their light output like the sun. Travelling to the bright star Mira in the autumn constellation of Cetus, the Whale, we find this red giant to be slowly changing its brightness. It pulsates in and out roughly every three hundred days and changes its brightness in the visible band by a factor of a hundred. At maximum it is visible to the naked eye, but at minimum only a telescope can pick it up. This variability was first noticed by a German clergyman, Fabricius, in 1596, fourteen years before the discovery of the telescope. He called the star Mira, the Latin word for wonderful. Many red giants show this same long-term variability, often with marked irregularity.

An unremarkable naked-eye star in the constellation of Cepheus, δ-Cephei, shows a much more regular but less dramatic variability in its light every 5.4 days. The star is in the giant, very luminous, stage before death and again we see the surface of the star oscillating in and out. 'Cepheid' variable stars like this play an important role for astronomers in measuring distances to stars. Normally it is very hard to tell whether a particular star is a very luminous star far away or a less powerful one near by. In the case of the cepheids it turns out that the greater the luminosity of the star the longer the period of pulsation. So if we spot a cepheid and measure its period of variability we know how powerful it is. And when we know the light is due to a candle rather than a lighthouse, or vice versa, we can tell how far away it is.

Variability with an entirely different cause appears in Algol 'the demon star', noticed by Geminiano Montanari in 1669. Regularly every 68 hours the star dims to one-third of its normal brightness for a few hours and then brightens up again. As we approach it we see that a faint companion star is eclipsing the primary star every two days. For a binary system to undergo eclipses is simply a matter of looking at it from the right angle, from a direction in the plane of the stars' orbits.

It is extremely unlikely that on our voyage we shall see one of those dramatic explosions that signal the final death of a massive star, a supernova. The ancient Chinese astronomers recorded several, including one in AD 1054 that has become famous as the origin of the 'Crab' nebula. If we travel a few hundred light-years towards the constellation of Taurus, the Bull, we can see the debris of this explosion, a weird crab-shaped cloud of filaments and nebulosity. We will encounter the Crab nebula (nebula was the name given by seventeenth-century astronomers to any fuzzy, fixed source of light that could be clearly distinguished from the point-like stars) on later voyages. Another very beautiful remnant of a supernova explosion lies in the constellation of Vela, where we see a more wispy, veil-like structure. And in the constellation of Cygnus, the Swan, we can make out a huge, faint, filamentary ring of visible light, relic of another supernova. The Vela and Cygnus supernovae exploded before the earliest historical records, but several supernovae have been seen in historical times. The first of the modern observational astronomers, Tycho Brahe, saw one in 1572, and his pupil, Johannes Kepler saw one in 1604. But none has been seen since then in the Milky Way, although astronomers believe that on average one should go off every fifty years somewhere in our Galaxy. Other supernovae which have been recorded in historical times, mostly by the Chinese, occurred in AD 185, 1006, and 1181. The supernova of 1006 was particularly spectacular. At its brightest it was a thousand times brighter than the brightest visible stars. A supernova in Cassiopeia, whose remains we will encounter on our radio voyage, is estimated to have exploded in about 1660, but nobody seems to have noticed it going off.

When a supernova explodes, the star flares up by a factor of a thousand million or more to its maximum brightness in a few days, then declines more slowly over a period of months. A supernova exploding in our Galaxy would normally become the brightest star in the sky for a week or so, visible perhaps even in the daytime for a few days.

The less violent phenomenon of the nova is, however, much more common: a star brightens by a factor of a million or so and then declines again over the next few months. There have been six bright novae in the present century, in 1901, 1918, 1925, 1935, 1942, and 1975. Although all originally too faint to be seen with the naked eye, they all became as bright as the brighter stars, and for a few days the 1918 nova was the brightest star in the sky. Novae seem to be the result of an exchange of matter between the stars of a binary system, one of which has to be a white dwarf. During a phase of the other star's life when it is expanding it dumps some of its material onto the surface of the white dwarf. This gas gets so hot that thermonuclear reactions are ignited and an explosion results, a kind of vast stellar H-bomb.

Another class of eruptive stars are called 'planetary nebulae', although they have no connection with planets. Three of the 107 objects in a very interesting list compiled in 1781 by the French comet-watcher, Charles Messier, are planetary nebulae and very beautiful they are too. He was interested in finding new comets and he did not want to be continually confused by other fuzzy or nebulous objects clearly visible with his telescope. So he made a list of nebulous objects that he was not interested in, objects to be avoided. Nowadays I think we would find the things in Messier's list even more interesting than comets. The list includes supernovae remnants (the Crab nebula is number 1 in the list, M1), star clusters (the Pleiades are M45), hot clouds of gas like the Orion nebula which we will meet shortly (M42), and whole star systems like our own Milky Way galaxy, but millions of light years away. The planetary nebulae consist of a circular ring or disc of bright light surrounding a bright, white star. They are caused by clouds of gas thrown off by stars late in their life. The gas can be seen lit up by the central star, which is on its way to the stellar graveyard to become a white dwarf. The cloud moves out slowly and contains quite a fraction of the star's mass. We shall encounter them again on our ultraviolet voyage.

Star clusters, gas clouds, and the Milky Way

So far we have explored the garden of stars and found several interesting species. We have seen stars at different stages of their life cycle, although the details of their birth and death will remain hidden until later voyages in other wavebands. Now we must travel into the depths of the Milky Way to try to find an overall pattern in this landscape of stars.

We are moving towards the landscape that is seen only with a large telescope. With his tiny telescope, Galileo made many discoveries: the phases of Venus, the four brightest moons of Jupiter, mountains on the moon, and that the Milky Way is made of stars. You can share his experience with a small telescope or pair of binoculars, although you must never look directly at the sun. Do what Galileo did instead and project the sun's image through the telescope onto a sheet of paper. William Herschel, a musician who became astronomer at the court of George III, was the first to build reasonably large telescopes and to explore the Milky Way systematically. Assisted by his devoted sister, Caroline, he found many thousands of nebulous or fuzzy objects and demonstrated the nature of some of them as clusters of stars. The largest telescope he built had a mirror 48 inches across and a tube 40 feet long. This telescope was very cumbersome and using it has been compared with trying to shave with a guillotine.

With a telescope we find that daylight is not the only obstacle to our voyage, for the night is not truly dark. Along the ecliptic or zodiacal plane we see a faint broad band of light, the zodiacal light which we have already noticed as sunlight reflected from solar systems dust. The air of our atmosphere glows faintly, particularly at certain wavelengths. And ruining many of the world's best telescope sites there is man himself and his city lights.

For our first clue to the large-scale distribution of stars we travel to the star group of the Pleiades, the Seven Sisters. If your eyesight is good you should be able to see with the naked eye seven or even more stars of this tight little group. As the group looms closer we see that in fact there are hundreds of stars in an irregular cluster. The most prominent stars are blue and luminous, the most numerous are stars like the sun and red dwarfs. All are still in their hydrogen-burning phase. Around some of the stars we see an irregular halo of light, where the star is lighting up some of the debris from the cloud of gas out of which the cluster was formed. The presence of the very massive luminous blue stars, up to fifty times as massive as the sun, shows that the cluster is relatively young. Such stars live out their lives in only a million years, compared with ten thousand million years for the sun, and hundreds of thousands of millions for the lowest mass red dwarfs. The Pleiades cluster was born less than a million years ago, and the earliest people on earth would have seen no Pleiades star group.

We travel now to the central 'star' of the sword of Orion where, 1500 light-years from earth, we seen an even younger group of four bright blue stars, the 'Trapezium', on the edge of a huge, patchy, shining cloud of gas many light-years across. This gas cloud, the Orion nebula, is a most beautiful sight in a big telescope, with its changes of colour from red to orange to green to yellow across the face of the nebula, the bright irregular lines that cross it, and the dark patches that eat into it. On a clear night the nebula looks fuzzy even to the naked eye. It is worth trying to find it for it has a special significance. When we travel here on our infrared voyage we will find that this is a place where new stars are being born.

Just bright enough to be seen by the naked eye, Messier's object number 22, in Sagittarius, turns out as we approach closer to be a dense spherical cluster of hundreds of thousands of stars. Clusters like this are called 'globular clusters' to distinguish them from the 'open' clusters like the Pleiades. We notice the overall red colour of the cluster in contrast to the blue Pleiades, where the many faint red stars do not contribute much of the total light. M22 is a much older cluster of stars, about ten thousand million years old. All the massive stars, which would have been luminous and blue while they were burning hydrogen, have gone to their deaths. The oldest stars burning hydrogen are not very different from the sun. There are many red giants in the cluster, the last stages of stars slightly more massive than the sun.

As we travel round the sky searching out these different types of cluster we start to get a picture of how the stars are distributed. The one thousand or so open clusters visible from earth are almost all very close to the plane of the Milky Way. The telescope shows this light to be made up of thousands of millions of stars. By counting the number of stars of different brightness in different directions on the sky, Herschel was able to tell that they are distributed in a disc with ourselves roughly midway between the two faces. This agreed with the theory put forward in 1755 by the German philosopher, Immanuel Kant, that the Milky Way is the light from a rotating disc of stars. The open clusters define a thin slice of meat in the sandwich of stars. Like the planets of the solar system strung out on the zodiac in their haze of zodiacal light, the open clusters too define a circle of the sky in the midst of the hazy Milky Way.

The globular clusters, on the other hand, of which more than a hundred are known, are far more spread out over the sky. There are slightly more of them near the Milky Way than elsewhere, but they have a more extended distribution than the light of the Milky Way. They form a nearly spherical halo about the disc of the Milky Way. The centre of this halo of globular clusters turns out to lie at a distance of about 30 000 light-years in the direction of Sagittarius. If we trace the Milky Way carefully round the whole sky — this would have to be done from somewhere in the southern hemisphere — we see that it bulges out in the direction of Sagittarius. This then is the direction of the centre of our star system, our Galaxy. The sun is in the suburbs of this metropolis, about two-thirds of the way out.

Suppose now we explore the detailed appearance of the Milky Way in our visible waveband. We find hundreds of examples of hot gas clouds illuminated by bright young blue stars, similar to the Orion nebula. Like the open clusters they lie close to the line where the Galaxy's plane of symmetry meets the Milky Way.

We see too that the outline of the Milky Way often seems to be eaten into by dark patches and lanes. Some of these dark patches can be seen with the naked eye on a very clear night. The most prominent, near the Southern Cross, has the appropriate name of the Coal Sack. What causes these? They seem to be dark clouds spread through the disc of the Galaxy which blot out the light from the stars behind, but what is in the clouds to blot out the light? A clue can be found from looking near the edges of these dark holes where a few stars begin to appear through the murk. These stars always turn out to be redder than average. Whatever cuts out the light has a worse effect on blue light than on red light. We remember the reddening of the sun at sunset and after a volcanic eruption and its cause in small particles of dust in the atmosphere. Are these dark clouds made of dust? The answer has to await our infrared voyage.

The landscape of galaxies

The stellar landscape has taken on its broad outline. Although much remains to be painted in, we seem to have incorporated all the wonders of the night sky. Yet in reality we have only taken the first step along the road of our voyage. For we have not mentioned the two most prominent nebulous objects on the sky, the Clouds of Magellan, noticed by the navigator as he crossed the southern oceans during 1521–2. We travel towards them, expecting them to be clusters of stars or clouds of gas in our Galaxy. Soon we realize that though we are leaving the disc of our Galaxy behind, they are still far off. Before we reach them we have travelled two hundred thousand light-years, three times the radius of our Galaxy. We find them to be irregularly shaped star systems, much larger than any cluster in the Galaxy, and containing many of the elements of our own Galaxy (open clusters, hot clouds of gas, etc.). They are full of young blue stars and clouds of gas, but not concentrated in a disc as in our own Galaxy, we see a halo of globular clusters around them. They are clearly independent galaxies, albeit much smaller than our Galaxy. The dominant mass of our Galaxy means that the Magellanic Clouds are our satellites, slowly orbiting us every few thousand million years.

We glance back towards our home. The huge disc of our Galaxy fills almost half the sky. The bright young blue stars and open clusters trace out huge spiral bands across the disc, like a catherine wheel. The centre of the Galaxy is marked by an enormous bright bulge of stars and clusters, hidden from us on earth by those dark clouds. With the biggest modern telescopes we could just pick out the sun towards the periphery of the disc, just inside one of the spiral arms. How the inhabitants of the Magellanic Clouds must admire our Galaxy and long to travel to it.

Some impression of this experience can be gained by travelling on towards a faint fuzzy spot in the constellation of Andromeda, number 31 in the list of Messier. Even with the naked eye, the medieval Arab astronomer Al-Sufi noticed the fuzziness of this image. In his *Book of the fixed stars*, published in AD 964, he calls it a 'little cloud'. The telescope reveals an oval-shaped cloud of stars, in the form of a circular disc tilted to our line of sight. Spiral 'arms' are clearly marked out by young blue stars and their associated hot gas clouds. This spiral structure in extragalactic nebulae was first seen by William Parsons (Lord Rosse) with the amazing 72-inch telescope which he completed in 1845. M31, the Andromeda nebula, is very similar to our Galaxy. It is ten times further from earth than the Magellanic Clouds, at two million light-years, and our home Galaxy at this distance begins to look remote.

The huge distance of M31 was first established by the American astronomer Edwin Hubble in 1927. He had abandoned possible careers first as a professional boxer, then as a lawyer, before taking up astronomy. The crucial step in establishing the distance of M31 and other near-by galaxies was the discovery of Cepheid variable stars in the spiral arms. By measuring their period of variation and using the known relationship between period and luminosity, Hubble could deduce the distance of the galaxy. His work on M31 and other near-by galaxies showed that the spiral nebulae were star systems far beyond the Milky Way and put an end to the long-standing debate about their nature. Spiral galaxies have three main components: a 'disc' of gas and stars, with spiral arms traced out by the young bright blue stars; a concentration of stars towards the centre of the disc, called the 'nucleus' of the galaxy; and a 'halo' of old stars and globular clusters enveloping the disc.

Like our Galaxy, M31 has two dwarf companions but these are very different from the Magellanic Clouds. They have a regular elliptical shape and contain only old red stars like the globular clusters of our Galaxy. There is no sign of a disc in these galaxies. They are the first examples we encounter of what are known as elliptical galaxies. Spiral and elliptical galaxies are the two main types of galaxy we find in the cosmic landscape, with an additional few per cent of irregular galaxies like the Megallanic Clouds and of peculiar galaxies that we shall meet later. Some of the 'ellipticals' turn out to be lens-shaped on closer examination and are called lenticulars. About a sixth of the spirals have a bright bar across their nucleus and their spiral arms start from the end of the bar instead of from the nucleus. These are called barred spirals. Spirals can be further classified according to how tightly wound up their spiral arms are, how prominent their nuclei are, and how blue they appear.

We see that there are altogether twenty or so galaxies in the space around M31 and our Galaxy but none are comparable in size to these two giants. This little flock of galaxies is called the 'Local Group' of galaxies. We have to travel ten million light-years to find the next galaxies, a group dominated by the spiral galaxy M81 and the large irregular cigar-shaped galaxy, M82. We have already travelled back in time past the origin of terrestrial man.

In every direction that we travel we find galaxies, stretching out as far as the eye can see with the largest telescopes. Almost all can be classified according to the simple scheme introduced by Hubble; i.e. into ellipticals, normal or barred spirals, or irregulars. Most galaxies are small groups each with a few to a few tens of members and vast spaces between each group. But spread across the constellation of Virgo, William Herschel's telescope revealed a huge cloud of thousands of galaxies — the 'Virgo' cluster — about fifty million light-years from earth. Most of the galaxies are ellipticals, but much larger than the Andromeda nebula's dwarf companions. A second huge cloud of galaxies can be seen ten times further away in the constellation of Coma Berenices, the Hair of Berenice. This has become known as the 'Coma' cluster.

We encounter these rich 'clusters' spread through the field of galaxies. They are the rare archipelagos in the ocean dotted with islands. To the limits of the world's largest telescopes we see a universe of galaxies and clusters of galaxies. However sitting on the earth we get at first a false impression. The first catalogues of nebulae by William Herschel and his son John, who continued the work of his father and his aunt, had fewer galaxies in areas of the sky near the plane of the Milky Way than in areas away from the Milky Way. This led many astronomers to assume that the 'spiral nebulae', as they were then known, were part of the Milky Way system. Only in the 1920s was Hubble able to demonstrate their very great distance. The reason for the apparent reduction in numbers of galaxies towards the Galactic plane is that small particles of dust are spread through the disc of the Milky Way, which scatter and absorb some of the light trying to reach the sun from outside of the Galaxy. Towards the Milky Way the light has to travel through a much longer column of dust and so is more likely to be absorbed or scattered. The galaxies in these directions then appear much fainter (and redder), so to any particular faintness limit there seem to be more galaxies towards the Galactic pole.

As we travel out to the faintest galaxies, which seem no more than the slightest smudges to the largest telescopes, we glimpse the cosmic landscape out to vast distances. The light from the most distant galaxies known set out many thousands of million years ago, before the earth was formed. As we travel back in time we see the galaxies further and further into their youth and we can trace their evolution with time. For the spiral galaxies we would see the proportion of their mass in the form of gas steadily increasing as we travel back into the past. The stars seem to form from the gas at a steady rate. At each epoch the light from the galaxies is dominated by the current generation of massive, luminous blue stars. For the elliptical galaxies, on the other hand, we do not see evidence of star formation even as far out as the limit of our vision. Instead we start to see light from stars of ever higher mass as we travel back in time, stars which would have been dead at the time we set off on our voyage. In their youth, ellipticals would start to look bluer and bluer as we saw the short-lived high-mass stars. It seems that ellipticals form all their stars at once, at their birth. The high-mass stars do not last long and at any epoch the light tends to be dominated by the most massive stars still alive at that epoch.

The spectrum

So far we have looked at the visible landscape much as the eye looks, paying attention only to the average colour of the light we see. What happens if we look at the visible wavelengths a bit more closely? I have mentioned that each kind of atom has its own characteristic wavelengths. Sodium atoms emit strong and pure orange radiation and mercury has a characteristic wavelength in the green, and these account for the colour of the street lamps which use these vapours. How does the cosmic landscape look when we pick one of these characteristic wavelengths and look only at the light of this wavelength?

Each atom has many characteristic wavelengths and there are many types of atom, so it would be tedious to examine each wavelength one by one. The astronomer uses an instrument called a spectrograph to look at them all at once. This consists basically of a prism which spreads visible light out into all its wavelengths or colours. Raindrops act like a prism on sunlight to give us the rainbow. Light from a star or galaxy is focused by a telescope onto a hole behind which lies the prism. The light passes through the prism onto a screen or photographic plate. Each wavelength present in the incoming light produces an image of the entrance hole at a slightly different position, because the glass of the prism bends or refracts each wavelength slightly differently. Usually the entrance hole is made to be a narrow slit, so the resulting image is a narrow line for incoming light of one particular wavelength. The images are recorded by photograph or electronically. The spread of colours produced by a prism is called the spectrum of the source. The rainbow is the sun's spectrum. The spectrograph shows thousands of dark lines across the sun's visible spectrum, where the atoms of the sun's surface layers have absorbed out their own characteristic wavelengths. By comparing these with laboratory spectra of different atoms, the relative amounts of the different atoms in the sun's surface layers can be found. The composition of the sun turns out to be rather similar to the earth, except that the earth has very little of the two most common elements on the sun, hydrogen and helium, and is also deficient in other lighter elements. Helium was first discovered as new element in the sun's spectrum, hence its name (from the Greek word for sun, *helios*). Only later was helium found in small traces in the earth's atmosphere. When Marie Curie discovered her new elements, radium and polonium, it was through their new and characteristic lines across the spectrum that she was able to prove that they really were new elements.

Which elements show strongly in a star's spectrum depends more on the temperature of the surface than on the composition. In hot blue stars we see the hydrogen and helium lines very strongly. Our sun's spectrum is dominated by lines of metals like iron and calcium. And very cool stars like red giants begin to show evidence of molecules like carbon monoxide and cyanogen, which can form at these lower temperatures. But if we compare stars with similar colours, and therefore similar temperatures, in different parts of our Galaxy, then we do notice differences. The metallic wavelengths show up more strongly in the stars of the disc than in the globular cluster stars of the halo of the Galaxy. We are seeing the formation of the elements in action. The older stars of the halo have practically no elements present except hydrogen and helium, but by the time the middle-aged stars like the sun were formed, the gas from which they were born is already well polluted with the elements formed in an older generation of stars now dead.

The landscape in motion

As we travel round the Galaxy in some chosen wavelength we

notice something interesting. The line does not always fall exactly at the expected point in the spectrum. The motion of the stars shifts the wavelength around through the Doppler shift. Stars moving towards us have the wavelength of their light shortened, while the wavelength of the light from receding stars is lengthened. Suddenly we see that the whole landscape is in motion. The stars are orbiting round the disc in a broad swirling motion, weaving in and out of the disc slightly, but with a mainly circular motion like the planets in the solar system. The stars of the halo have a more agitated appearance, diving through the centre of the Galaxy, out the other side until gravity brings them to rest, and then back again. We guess that these old stars were formed early in the life of the Galaxy, when the huge gas cloud which was to become the Galaxy was still collapsing together. The stars retain much of the energy of motion of the gas they were born from. The disc, on the other hand, seems to have been formed out of the gas left over from the halo generation of stars, and to have settled into a more ordered circular motion before stars started forming there.

This scenario seems to apply to all spiral galaxies. The halo stars form during the collapse of the gas cloud which is condensing to become a galaxy. The spin of the cloud forces most of the remaining gas into a rotating disc, though some manages to form a large concentration at the centre, the nucleus. Star formation then starts in the disc and continues till the present day, with less and less gas left as time goes by. The elliptical galaxies have no disc and no strong rotation so they seem to have done their formation all at once during the collapse phase.

When we look at the spectrum of a distant galaxy we are seeing the average of the light from all the stars, so we see a mixture of the characteristics of different temperatures, though usually with some predominant type of spectrum. The red ellipticals, for example, show the typical atomic wavelengths of the cooler stars which dominate their light. Any particular atomic wavelength is spread out by the motions of the stars in the galaxy, to give rather broad lines in the spectrum.

By studying how fast the stars are moving around in different parts of a galaxy we can get an idea of how strong gravity is there and hence how heavy the galaxy is. In this way we find that there are dwarf galaxies of a hundred million times the mass of the sun and giant galaxies of a million million times the sun's mass. Our

Galaxy is quite a big one, about two hundred thousand million times the mass of the sun. The central regions of a galaxy, the nucleus, tends to contain a few per cent of the galaxy's mass within a region a thousand light-years or so across.

This technique of using star motions to measure gravity has recently been applied to the central regions of a prominent elliptical galaxy in the Virgo cluster of galaxies, number 87 in Messier's list, with surprising results. The motions near the centre are much faster than expected, showing that a substantial mass, several thousand million times the mass of the sun, is concentrated within a hundred light-years of the centre of the galaxy. If we travel to these central regions, we find there is very little light coming from them and so this mass cannot be in the form of normal stars. However we do notice something unusual about the nucleus of M87. Short-exposure photographs show a jet of light sticking out of the nucleus, thousands of light-years long, drowned out by starlight on a normal photograph of the galaxy.

To find a phenomenon closer to our experience of a similar origin, we have to turn in a most unexpected direction, to the aurora borealis, the 'northern lights', or the aurora australis, the 'southern lights'. In the extravagant words of the Norwegian polar explorer F. Nansen:

The aurora shakes over the vault of the heaven its veil of glittering silver — changing now to yellow, now to green, now to red. It spreads, it contracts again, in restless change, next it breaks into waving, many folded bands of shining silver, over which shoot billions of glittering rays: and then the glory vanishes. Presently it shimmers in tongues of flame over the very zenith: and then again it shoots a bright ray up from the horizon, until the whole melts away in the moonlight, and it as though one heard the sigh of a departing spirit. Here and there are left a few waving streamers of light, vague as a foreboding — they are the dust from the aurora's glittering coat.

Like the jet in M87, this inspiring radiation has its origin in electrons moving very fast, at speeds close to that of light, spiralling along a magnetic field. In the case of the aurorae, the electrons are accelerated to their huge speeds in active regions on the sun, for example in a solar flare. When they reach the earth they spiral along the earth's magnetic field lines until they crash into the earth's atmosphere, where they heat the atmospheric gases to a state of glowing incandescence. In the jet of M87 we probably see radiation directly from the fast-moving electrons as they spiral along the magnetic field: 'synchrotron' radiation. This radiation in M87 hints at violent events in M87's dark compact nucleus. This mystery will become clearer after our radio and X-ray voyages.

The redshift and the expanding universe

As we travel towards the more distant galaxies, we notice a most unexpected change in the spectrum of the galaxy. We can recognize the pattern of atomic wavelengths, but they do not fall at the same wavelengths as their laboratory counterparts. The galaxy wavelengths have all been lengthened, their frequencies reduced, each by the same factor. The simplest explanation is that the galaxies are moving and the wavelengths are being Doppler shifted. In the visible spectrum we see that the lines are shifted towards the red end. This 'redshift' of the galaxy light applies not only to the spectral lines but also to the thermal radiation of the hot stellar surfaces, though this is harder to spot. We are seeing the motion of the galaxies and they all seem to be receding from us. There is a precise relationship between how fast the galaxy is moving away from us and how far away it is: the further away the galaxy is from earth, the faster its speed of recession, and the two are exactly proportional to each other. If we look at two galaxies one twice as far away as the other we find the one is receding twice as fast as the other. This is called the 'velocity–distance' or Hubble law, after Edwin Hubble, who discovered it in 1929. The whole universe is expanding as if we are seeing the sparks from some immense explosion. At first we seem to be the centre of this explosion since the galaxies recede from us whichever direction we look in. But as we travel to a distant galaxy and look around us, we see almost the same picture. The only difference is that because we have travelled back in time, the galaxies are crowded in closer together. Was there a time when they were all touching? We shall have to await our microwave voyage to find any clue to this.

We can, however, try to guess what will happen in the future. The expansion of the universe does not go unresisted. The gravitational attraction between the galaxies tries to halt the expansion. Whether it will eventually succeed depends on the average amount of matter in any volume of the universe and on the rate of expansion. Present estimates suggest that the gravitational attraction between the galaxies will not be enough to halt the expansion, so the universe will just keep on expanding for ever.

One of the questions that we ask ourselves as we look up at the night sky is: Does the universe go on for ever? Is it infinite in extent? Now I said just now that the universe does not have a centre, so you may feel that settles the question. Surely if the universe were finite in extent, like the earth is, then it would have a definite centre. And if it was finite, like the earth, then there would be an inside and an outside, and how can there be any-where outside the universe? And wouldn't the universe have an edge, a boundary, which just sounds nonsense? These arguments were used by the Roman poet Lucretius in the first century AD to argue for an infinite universe. They sound pretty convincing, but they do not reckon with the possibility of curved space. It turns out that in a curved space you can have a finite universe which doesn't have a boundary and doesn't have an outside and everywhere does seem to be at the centre. In such a universe if you travel off in any particular direction and keep on going in a straight line for a very long time then you will eventually arrive back at your starting point from the opposite direction.

What kind of universe are we in and is space curved? No direct progress has been made in answering this. In the simplest possible models of the expanding universe, however, there happens to be a connection between whether the universe is finite or infinite, and whether it will keep on expanding for ever. If gravity is going to halt the expansion and make the collapse together again into a second, imploding Big Bang, then the models say that the universe is finite. If, as seems to be the case, the universe will keep on expanding for ever, then it ought to be infinite. Actually space is curved in this case too, but instead of being curved in a way that makes the universe close up in a ball, space is curved in a way that just keeps opening the universe out more and more.

We have not yet completed our voyage in the visible wave-band. Other voyages will reveal new objects, new phenomena, which will send us back to this familiar band. But we have completed the voyage of the old astronomy. We focus on the most distant galaxy discovered purely by optical telescopes, the brightest in the cluster denoted '1304+29'. We have voyaged back in time almost ten thousand million years. Looking back at our own utterly distant Galaxy, the sun and earth have not yet formed and perhaps no life exists yet in the universe. Where are the ingredients that will make life possible? Is it inevitable that life and intelligence will arise? At the end of our voyages we will have no certain answers to these questions. But at least we know that life has arisen here on earth, that we exist and are capable of this voyage through the cosmic landscape.

3
SECOND VOYAGE
The radio landscape

What are radio waves?

We return to earth from the limits of the known universe and equip ourselves for our second voyage. We are to travel in an entirely new waveband, that of radio. Instead of the eye or the photographic plate, we detect radio waves with specially designed radio receivers. Radio waves were first discovered by Heinrich Hertz in 1888 and first used practically by Guglielmo Marconi in England in 1895. Their breakthroughs followed the sensational recognition by James Clerk Maxwell, in 1864, that light consisted of nothing more than an oscillating electromagnetic field. At any point in space an electric charge will be pushed in the direction of the local electric field, if there is one. A magnet will line itself up in the direction of the prevailing magnetic force field. The earth itself acts like a huge bar magnet and at any point on or near earth's surface, a compass needle, which is nothing more than a small magnet, will align itself in the direction of the earth's magnetic field. Now when a light wave passes all that we notice is a trembling of the electric and magnetic fields and then stillness again. That trembling is the light wave.

The wave pattern travels along at the speed of light and like the waves on the sea it carries energy with it. This energy can illuminate our visible landscape if the waves have visible wavelengths, warm us if they have infrared wavelengths, or make a radio receiver work if they have radio wavelengths. When we try to make oscillating magnetic and electric fields in the laboratory, it is natural that our apparatus will tend to have a scale of centimetres or metres, which is why the first efforts to synthesize light led to the discovery of a new kind of light, radio.

How does a radio work? When radio waves fall on a long piece of wire connected to the ground, a tiny electric current flows back and forth along the wire in time with the wave, driven by the varying electric field of the wave. The wire will respond much more strongly to radio waves of wavelength about the length of the wire. To tap this electrical energy we need to connect to the wire an electrical circuit with a 'rectifier' in it. A rectifier is a device which only allows the current to flow in one direction along the circuit. We then use this current to power a pair of headphones. The simple 'crystal' set, which gives a very adequate performance in the medium waveband, is not much more complicated than this. A long piece of wire, the 'aerial', is connected to the ground to respond to the radio waves. The current flowing to and fro along the aerial is drawn off into a simple electrical circuit which can be 'tuned' to respond to waves of only one particular wavelength. The rectification is provided by a silicon or germanium crystal diode. Your pocket transistor set elaborates this circuit by replacing the crystal with a transistor which also amplifies the signal crossing it and by providing subsequent stages of amplification to feed a loudspeaker. The beauty of the crystal set is that it powers the headphones directly with the energy of the incoming radio wave, whereas transistors or other types of diode require a source of electricity.

The receivers of radio astronomy are related to the pocket transistor much as a Formula One racing car is related to the family car: they are real specialist jobs, very sensitive, and highly tuned. They have to be to filter out the local human din and pick up the whispers from the cosmic landscape.

The birth of radio astronomy

The birth of radio astronomy was also the birth of the new astronomy, the astronomy of the invisible wavelengths which is the subject of this book. It is true that ground-based astronomy had long pushed out in wavelength a little way beyond the visible band, into the near infrared and near ultraviolet, but the explosion of knowledge in these bands was to follow that in the radio band. The new astronomy began quietly with a routine investigation of radio 'static' interference at a wavelength of 15 metres by Karl Jansky, at the Bell Telephone Laboratories during 1930–3. He built a rotatable aerial array 30 metres long and 4 metres high, mounted on four wheels taken from a Model T Ford, nicknamed the 'merry-go-round'. He found that part of the static was due to thunderstorms; intermittent crashes due to near-by ones and a steadier weaker noise due to many distant storms. But there remained a steady weak hiss from a direction which moved around the sky a little each day. This direction turned out to be the constellation of Sagittarius. Jansky had measured radio emission from the Milky Way.

Few astronomers took any notice of his results and Jansky himself had to return to his normal telecommunication assignments. The development of radio astronomy would have come to a halt but for a remarkable character, the American amateur astronomer Grote Reber. Fascinated by Jansky's discovery he decided to build a radio telescope in his back garden. To collect the radio waves more efficiently he constructed a large metallic dish in the shape of a parabola. This reflected the radio waves and focused them onto his aerial. With this magnificent 31-foot parabolic reflector, which must have amazed his neighbours, he used to observe the radio sky between midnight and 6 a.m., before driving to work for a radio company during the day. Between 1938 and 1944 he produced an admirable radio map of the Milky Way at a wavelength of 1.87 metres. American astronomy, dominated by the optical astronomy made possible by the large optical telescopes, failed at first to build on the work of these two pioneers.

Meanwhile in wartime Britain, the third and most influential of the three pioneers of radio astronomy, J. H. Hey, had started the investigation which led to several astronomical discoveries and to the formation of radio astronomy as a branch of science. Neither an astronomer nor a radiophysicist, he had taken a job in 1942 as leader of a small group with the task of investigating enemy radar jamming for the army. The enemy turned out to be the sun. (At Bell Telephone Laboratories, G. C. Southworth independently detected radio emission from the sun during 1942.) After the end of the war, Hey's group discovered radio emission from meteor trails and, far more significantly, the first discrete radio source. This was Cygnus A, in the constellation of Cygnus, the Swan.

When the first sources of radio emission distinct from the general radiation of the Milky Way were discovered, their positions were so poorly known that they were just given the name of the constellation they happened to be in the direction of, with a letter. Other prominent radio sources found soon after Cygnus A were called Cassiopeia A, Taurus A (this turned out to be the Crab nebula), and Virgo A (which was in fact M87).

The work of Hey's group led directly to the formation of the two major British radio observatories. Bernard Lovell and Martin

Ryle had both worked on airborne radar during the war. Afterwards they returned to their universities, Manchester and Cambridge, and started work on radio astronomy with the receivers and aerials no longer needed for radar. Lovell worked on the radar echoes from meteor trails, with some initial assistance from Hey, and later conceived the idea of the 250-foot Jodrell Bank telescope in order to detect radio echoes from cosmic rays. Ryle decided to observe the sun in more detail and started to develop 'interferometer' radio telescopes, where the output from two separate small dishes is combined to give the same ability to see fine detail, or 'resolving' power, as a large dish.

In Australia the development of radio astronomy also began during these post-war years, under J. L. Pawsey, and their first observations were also of the sun, using a different type of interferometer to Ryle. Radio waves reflected from the sea were combined with those observed directly by a cliff-top radio telescope. Thus began the rivalry between Cambridge and Australian radio astronomers which was to continue for many years.

The thirty years since the work of Hey's group have seen phenomenal progress, nowhere more strikingly than in the telescopes themselves. They are among the most beautiful engineering creations of these decades. Jodrell Bank and Cambridge, England; Parkes, Australia; Owens Valley and Hat Creek, California; Green Bank, West Virginia; Westerbork, Netherlands; Medocina, Italy; Nancay, France; Effelberg, Germany; Arecibo, Puerto Rico; Ooty, India; Soccoro, New Mexico. Any of these places, and many others, are worth a pilgrimage to see the delicate traceries of metal gazing at the sky. They have to be so large because what counts when you are trying to resolve small detail in the landscape is how big your light collector is in terms of the wavelength you are using. The pupil of the eye is about six thousand visual wavelengths across and can distinguish two points separated by about one minute of arc, which is the angle made by a sixpence or dime about half a mile away. At 178 MHz, or 1.7 metres wavelength in the radio band, a similar resolving power would need a telescope 10 kilometres across. A single dish this size would be too unwieldy and colossally expensive so radio astronomers use several much smaller dishes strung out in a long line. When the signals from these are combined and the earth's rotation is used, a very large telescope can be laboriously simulated. The largest fully steerable dish is at Effelberg and is

100 metres in diameter. At Arecibo a natural valley has been used to make a fixed dish 305 metres across which can observe objects as they cross above it.

The earth's environment

Not all radio frequencies from the cosmic landscape reach the earth. The lowest frequencies (longest wavelengths) do not penetrate to the earth, for above the earth's atmosphere is a layer of electrically charged atomic particles (ions and electrons) called the 'ionosphere' which reflects low-frequency waves away. It also incidentally assists long-wave communication by reflecting mankind's transmissions back to earth. This is how radio waves, which travel in straight lines, can be sent round the curved earth.

The ionosphere introduces us to an important state of matter not familiar to us in the terrestrial environment, ionized gas. This is gas where the atoms have lost one or more of their electrons. This can happen through collisions between atoms when the gas is very hot or because the atoms of the gas have been bombarded by very energetic photons. The loss of the electrons, which carry a negative charge, leaves the residual atoms with a positive charge. They are called 'ions' in this state. The positive ions and the negative electrons rush around freely in the gas and respond to any electric field that is present. An ionized gas does not allow low-frequency radiation through at all because the ponderously varying electric field of the electromagnetic wave is immediately neutralized by the rapid motions of the ions and electrons. The free electrons in the gas also act as scatterers of radiation, but how effective this scattering is depends on how much gas there is.

As well as absorbing and scattering light, an ionized gas is also a source of radiation through a mechanism called 'free–free' radiation. A free electron rushing around through the gas some-times passes rather close to an ion. When it does so it comes under the influence of the electrical attraction of the ion (unlike charges attract each other). During this period of close encounter with the ion, the electron loses some energy in the form of photons. This radiation emerges from the gas with a very wide range of frequencies. It extends all the way from the radio band to the visible band, and to higher frequencies still if the gas is very hot. This is the signature of an ionized gas cloud if it is not so dense and opaque that we cannot see right through it. If the cloud does become very dense and opaque, the distribution of radiated energy with wavelength (the spectrum of the cloud) goes over to that of a hot body at the temperature of the gas (a thermal spectrum).

One of the few examples of ionized gas on earth is the track of a lightning flash. The discharge of electricity from a thundercloud to the ground ionizes the air all along its path for a brief instant. The sun is an ionized gas cloud and in fact much of the matter in the universe is probably in an ionized state. The earth's ionosphere, mentioned above as a barrier to radio astronomy, is formed by ultraviolet and X-ray photons from the sun bombarding the top of the earth's atmosphere, stripping electrons from atoms and molecules of air. The ionosphere limits ground-based radio astronomy to frequencies above about 10 MHz. Karl Jansky made his historic observations at 20 MHz, close to the limit. Had he chosen a somewhat lower frequency he would not have been able to detect the Milky Way. Satellites in high orbits (above 6000 kilometres) can extend this limit down to 300 kHz and at 100 000 kilometres from earth frequencies as low as 25 kHz (a wavelength of about 12 kilometres) could be detected from the cosmic landscape. However the interplanetary gas reaching out from the sun past the earth prevents lower frequencies than this arriving at the vicinity of the earth.

At the very-high-frequency (very-short-wavelength) end of the radio band, the molecules of gas in the earth's atmosphere start to absorb the waves. We shall treat these very high radio frequencies as a separate band, the microwave, and make them the medium of our final voyage.

The sun and Jupiter

It is time to depart. As we leave earth's atmosphere, we see a faint glow of free–free radiation from the ionosphere. Further out, at about 30 000 to 100 000 kilometres from earth's surface we see the Van Allen 'radiation belts', doughnut-shaped zones of charged particles girdling the earth's equator. The Van Allen belts are formed by electrons and protons arriving from the sun at very high velocities—close to the speed of light—and becoming trapped by the earth's magnetic field. When a charged particle runs into a magnetic field, it experiences a force that is perpendicular both to its direction of motion and to the direction of the magnetic field. The particle is therefore forced into a circular or spiralling motion around the magnetic field. Many of the charged particles arriving from the sun find themselves channelled into a zone girdling the earth's equator. They spiral backwards and forwards along the magnetic field lines but they are trapped. Now, when the particles move close to the speed of light an effect of the theory of relativity comes into play and they radiate strongly as they spiral along the magnetic field. This radiation is called synchrotron radiation and we encountered it in our visible voyage in the jet in the nucleus of M87. Particles moving at speeds close to that of light are called 'relativistic' particles. In the radiation belts the relativistic particles spiral to and fro frenetically, radiating synchrotron radio emission. The discovery in 1958 of the radiation belts by the group led by James Van Allen of the University of Iowa, using the *Explorer I* satellite, marked the birth of a whole new area of space science, that of magnetospheric physics. This is the study of the region of space controlled and shielded by earth's magnetic field, the 'magnetosphere'.

We see that a stream of new relativistic particles rushes towards the earth from the sun in a great wind, some to be trapped in the radiation belts, most to be deflected round the earth by its magnetic field, a few to travel on towards the unprotected polar regions to dissipate their energy in the magnificent spectacle of the aurorae. These wonderful displays of radiation generated by particles moving close to the speed of light give us a vivid feel for the way the solar 'wind' buffets the earth. This wind of gas and fast particles was discovered in the same year as the Van Allen belts and has completely changed our ideas about the space between the sun and the earth. Gas is continuously accelerated from the surface of the sun and fills interplanetary space with an ever-expanding and accelerating wind. The surface of the moon is directly blasted by this wind, but the earth's magnetic field creates a cavity round which the wind is swept, called the 'magnetopause'.

Perhaps we are travelling when the sun is disturbed by spots and flares on its surface. The solar wind becomes a gale, the radiation belts buckle to and fro, the ionosphere is buffeted about. Compass needles around the earth go haywire as a magnetic storm rages. Radio communications are wiped out and the radio band is flooded with bursts of radio emission from the direction of the sun. These radio bursts are caused by electrons moving close to the speed of light, accelerated in the flare event.

Yet when the storm is passed the sun has a pallid and bloated appearance in our band. As we travel towards the sun we find that its radio emission comes from far beyond the surface defined by visible light, from a zone known as the corona, after its crown-like appearance in visible light during eclipses. In visible light the disc of the sun is evenly bright, but in radio light the sun is brighter towards its rim than its centre, and is brighter towards the equator than towards the poles. The sun is not specially dazzling to our radio eyes, and we can hardly pick out any of the other familiar stars at all.

As we travel on outwards, we notice that only Jupiter stands out amongst the solar system planets. Jupiter's surprisingly strong radio emission was discovered accidentally by two US astronomers, B. F. Burke and K. L. Franklin, in 1955. They found inermittent bursts at low radio frequencies (around 20 MHz) which may be generated by lightning flashes in Jupiter's atmosphere and ionosphere. Curiously the brightness of these bursts of radio emission is affected by the passage of Jupiter's satellite, Io. At higher radio frequencies (wavelengths of tens of centimetres) some of the heat from the cool disc of the planet is seen (most comes out at infrared wavelengths), but also strong synchrotron radiation from relativistic particles trapped in Jupiter's radiation belts. Jupiter's magnetic field is ten times stronger than the earth's, its magnetosphere is a hundred times larger, and its radiation belts far more impressive than the Van Allen belts round earth.

The other planets can be seen only dimly in the radio band. Their temperatures range from warm to cool and most of their heat, or thermal energy, is radiated in the infrared, so we shall see them more clearly on our infrared voyage. However part of their thermal radiation is emitted as radio waves and so we do see a faint glow from them.

As we look round the radio sky, the stars have vanished and the brightest radio sources are in directions we had hardly noticed at all in the visible band.

The Milky Way

Dominating the sky is the radio emission from the Milky Way, which spreads over almost the whole sky at the lower radio frequencies. As we trace the radio photons back to their source we find they are being radiated by relativistic electrons spiralling in the Galaxy's large-scale magnetic field. These same electrons are among those atomic particles striking the earth which earlier in this century were given the name 'cosmic ray'. Cosmic rays consist of electrons, protons, and ions of other atoms (the ion of hydrogen is the proton) all moving relativistically. The most energetic of these have speeds within one part in a million million of the speed of light. Where have these cosmic messengers come from? Most come from the sun as part of the solar wind. The gaseous part of the solar wind streams past the magnetopause, the cavity in the wind created by the earth's magnetic field, while the lower energy relativistic particles are trapped in the radiation belts and the higher energy particles stream onto the earth as cosmic rays. The remainder of the cosmic rays come from much further away and permeate the whole of our Galaxy. The Galactic cosmic-ray electrons are the particles responsible for the radio emission from the Milky Way. These probably originate in supernovae remnants, more of which shortly. A few of the highest energy cosmic rays may come from outside the Galaxy, from explosions on a galactic scale we have not yet encountered in the cosmic landscape.

When we turn our attention in more detail towards the plane of the Milky Way, we find bright extended sources strung out along the Galactic plane like beads on a necklace. They stand out as very brilliant patches against the Milky Way at high radio frequencies, but at low radio frequencies they appear as dark absorbing patches. As we travel towards them we realize that many of them are the clouds of hot gas, heated by young, massive blue stars, that we noticed on our visible voyage. Both the radio and visible radiations are free–free emission from the ionized gas. These blue stars are actually radiating the bulk of their energy in the ultraviolet, as we shall see on our next voyage, and it is these energetic ultraviolet photons from the star which strip the electrons off the atoms of gas and ionize it. As we trace these clouds round the Galaxy we are witnessing the locations of recent star formation, for the associated stars do not live for more than ten million years or so. The clouds are not spread evenly through the disc of our Galaxy. Instead they are concentrated in a spiral pattern which can be followed right round the disc. If you trace out the Milky Way on the sky from Cassiopeia through Taurus to near Orion, you are looking at the spiral arm nearest to earth, only a few hundred light-years away.

If we make our voyage in a very particular radio wavelength, 21 centimetres, we suddenly see a different and much more intense sky. The reason is that atomic hydrogen has a characteristic wavelength here. Spiral arms stand out very strongly as long chains of clouds of cold gas. We realize that these are the reservoir of material for new stars and the spiral arms are the places where the clouds contract to form stars. Not till our infrared voyage will we actually see the birth of a new star. About 1 per cent of the mass of our Galaxy remains in the form of atomic gas, waiting for its turn to form into stars, but the rate of formation is rather slow now. The gas is being replenished too, by the gas being ejected from the sun in the solar wind and from other stars in their similar 'stellar winds', and by the more dramatic events at the end of stars' lives. The death throes of stars and the steady loss of mass throughout a star's lifetime are phenomena we shall encounter on other voyages.

Supernova remnants and the metronome of the pulsars

Let us travel now towards the brightest discrete source of metre-wavelength radio waves in the sky, brighter even than the sun, Cassiopeia A. As we approach we see that it consists of a spherical husk or shell 10 light-years across, emitting intense synchrotron radiation. If we could wait around a while we would see the shell expand outwards to become even larger, like the mushroom cloud of an H-bomb. It is the relic of a vast explosion about three hundred years ago, when a star at least ten times bigger than the sun greedily came to the end of its nuclear fuel and destroyed itself as a supernova. This particular star had been losing mass steadily and prolifically for a hundred thousand years before the catastrophe.

Altogether about a hundred and twenty of these radio-emitting remnants of supernovae are known in our Galaxy, including remnants of the supernovae seen in visible light by the Chinese and European astronomers in historical times. These explosions spray the interstellar gas with the products of the nuclear reactions that have been taking place in the star. Gradually the concentration of elements heavier than hydrogen and helium increases, though not to a very high level. Earth, composed predominantly of these heavier elements, is an untypical location in the cosmos, where on average hydrogen and helium account for 98 per cent or more of all matter at the present epoch.

We move across to another of these large shells of radio radiation, in the constellation of Taurus. In the visible band we saw something interesting here, an irregular cloud of light with limbs and filaments sticking out in all directions with the nickname of the 'Crab'. This is the relic of a supernova explosion, noticed by Chinese astronomers as what they called a 'guest star' in AD 1054. In the radio band too we notice something interesting. From a point in the centre of the Crab nebula, the position of a faint star in fact, we see the radio signals flickering with great regularity every thirty-three thousandths of a second — a 'pulsar', short for pulsating radio source. The radio pulses repeat themselves with such clock-like precision that when the first pulsar was discovered by the British radio astronomers Tony Hewish and Jocelyn Bell in 1967, they seriously considered the possibility that it was a radio transmission from another civilization. However this soon turned out to be improbable because there was no sign of the Doppler shift expected due to the motion of the hypothetical planet round its parent star. Hundreds of pulsars have now been found round the Milky Way, most having pulsation periods between one-quarter of a second and 2 seconds. Only the two youngest, the Crab pulsar and another in the constellation of Vela, are in recognizable supernovae remnants. The remnants of the supernovae explosions which gave birth to the other pulsars must have faded away long ago.

As we travel towards the Crab pulsar we see how tiny it is compared with the Crab nebula itself, only a few hundred miles across compared with the millions of millions for the nebula. The beam that gives the regular pulse we had noticed sweeps round like a lighthouse beam, thirty times a second. In our band we see only a cloud of charged particles and magnetic field, but deep inside this cloud, only ten miles across, lie the remains of the core of the star that exploded. It has collapsed to an incredible density — a matchboxful weighs a ton — and the surface has become a crystalline solid; it is called a neutron star. If we could not feel the powerful gravity we might mistake a neutron star for an asteroid, with its solid dark surface only ten miles or so across. Yet packed into this tiny volume is a mass equal to that of our whole sun. The pulsar phenomenon is associated with the neutron star's magnetosphere. Neutron stars have phenomenally strong magnetic fields, a million million times stronger than that of the earth, so it is not surprising that the neutron star's magnetosphere is altogether a more spectacular affair. The period of the pulsar's pulsation is presumed to be the neutron star's rotation period — it is rotating millions of times faster than the sun's stately once per thirty days. How the rotating magnetosphere of the neutron star produces the radio pulse is still something of a mystery.

A neutron star is the second kind of death throe for a star. On our visible voyage we encountered an example of the first kind, the white dwarf companion of Sirius. A white dwarf is not quite as compact as a neutron star, perhaps a hundred times larger, and we can still see the star's gas cooling down. The surface of a neutron star is invisible and quiet in the radio, for it is its magnetosphere far away from the star's surface that creates the pulsating radio source. Only on our X-ray voyage will we come close to the neutron star itself and also to the third possible fate for a dead star, a black hole.

The violence of the radio galaxies

We leave our Galaxy behind and search out our neighbouring galaxies. The Andromeda nebula is naturally very similar to our own Galaxy, with synchrotron emission from relativistic electrons spiralling in the disc of the galaxy and 21-centimetre atomic hydrogen radiation from cold clouds along the spiral arms. The latter are also traced out by radio-emitting hot clouds lit up by young stars. The main surprise perhaps is the extent of the cold atomic hydrogen, spreading to almost twice the size of the starlit disc. The irregular galaxies like our companions the Magellanic Clouds show abundant atomic hydrogen and the proportion of the total galaxy mass in the form of gas may jump to 50 per cent or more in this type of galaxy. On the other hand the two elliptical companions of M31 show no sign of radio emission at all, nor do other near-by elliptical galaxies, either from synchrotron radiation, from hot ionized gas clouds, or from cold clouds of atomic hydrogen. We recall also that elliptical galaxies contain only old red stars. Thus there is a close connection between seeing young, bright blue stars in a galaxy, as we do in spiral and irregular galaxies, and seeing gas clouds from which new stars can form. In elliptical galaxies star formation was over long ago, whereas in spiral and irregular galaxies it continues today. Also since all the massive stars in ellipticals completed their life cycle long ago when the galaxy was young, there are almost no supernovae today to provide the cosmic-ray electrons needed to create synchrotron radio emission. A patrol of hundreds of bright galaxies of all types is maintained by some astronomers of the visible band to watch for supernovae. For a few days a supernova may stand out brighter than the whole of the starlight from the rest of the galaxy. Many hundreds have been seen and this is a consolation for our failure to see one in our part of the Milky Way since 1572.

The radio landscape seems to be taking an ordered shape. Yet we have only to travel a little further to find the first example of the phenomenon which made radio astronomy famous and which signalled the emergence of the new astronomy. In the southern constellation of Centaurus, after a journey of 15 million years back in time we find a source thousands of times more luminous in radio waves than our whole galaxy, called Centaurus A. As we approach we find the emission is from two huge clouds of relativistic electrons a million light-years across and several million light-years apart. To understand what this double radio source means we need some help from another wavelength band. What would the visible band have shown? Midway between the positions of the two radio clouds is an unusual looking giant elliptical galaxy, crossed by a swathe of obscuring dust. As the galaxy is only a hundred thousand light-years across, can we be sure there is any connection with the much larger radio lobes spilling out on either side? As we focus in on the position of the centre of the visible galaxy with our radio eyes we find a second pair of radio sources on a much smaller scale, within the area covered by the stars of the galaxy. These sources are precisely aligned with the more powerful extended lobes, and also neatly straddle the very centre of the visible galaxy, so we can feel confident that all the radio emission does indeed originate in the galaxy. And at the very heart of the galaxy, midway between the two inner radio lobes, we find a still weaker compact radio source, more prominent at higher radio frequencies. This seems to be the point of origin of the radio emission, deep in the galaxy's nucleus.

When we work out the total number of relativistic electrons

needed to provide the radio power of these vast outer sources and assess the amount of energy stored in the clouds in the form of relativistic particles and in the magnetic field, we find a huge total energy. In fact if we compare this with the total energy of the matter in the nucleus of the galaxy, we find that creating the radio sources must have consumed a substantial fraction of the entire available energy budget of the galaxy nucleus. So colossal are the energy demands of the radio sources that an origin in the relatively inefficient nuclear processes in stars seems ruled out. Only about one per cent of the matter of a star is turned into the energy of starlight by nuclear reactions. As well as the mystery of how the energy that emerges in the radio band is generated, we have to understand why this energy is released so far out from the centre of the galaxy. If the galactic nucleus had undergone some vast supernova-like explosion we might have expected the source to be in a spherical cloud centred on the galaxy nucleus itself.

As we gaze out at the radio sky, the majority of the brightest sources away from the plane of the Milky Way turn out to be other examples of this phenomenon. The familiar stars of the visible sky are undetectable in radio, a few only becoming faintly detectable when they undergo violent flares. Even the normal galaxies of the Local Group, and of other near-by groups and clusters of galaxies, are extremely faint. The majority of the brightest radio sources away from the Milky Way are rather distant galaxies, the so-called 'radio galaxies', singled out for our attention by their powerful radio emission. Radio galaxies seem always to be elliptical galaxies, often very large and massive ones.

We travel on, to the core of that nearest large cluster of galaxies, the Virgo cluster. Centred on the second brightest galaxy in the cluster, the spherical galaxy M87, is a strong highly elongated double radio source. The lobe on one side coincides with the jet of visible light we saw earlier sticking out of the nucleus.

Before we try to relate M87's powerful radio output to the dark compact mass we found in its nucleus in the visible band, let us look at some other examples of the radio-galaxy phenomenon. We voyage on further to a cluster of galaxies a thousand million light-years away in the direction of Cygnus, the Swan. The radio source, Cygnus A, centred on one of the galaxies in the cluster, is one of the most luminous radio galaxies known. The radio structure, as revealed by the Cambridge 5-kilometre telescope, is very suggestive of what is happening in these radio galaxies. At first we see the usual double structure, but as we approach the galaxy we see that inside each of the huge lobes, at the far end of them away from the galaxy, there is a powerful, very small bright spot of radio emission. These two bright spots lie exactly on a straight line through the nucleus of the galaxy and strongly suggest that the radio lobes have been formed by a double beam of energy fired out from the nucleus and then impinging on gas spread through the space around the galaxy. The source looks like the result of a fire hose spraying on a wall. The beam could be composed of relativistic particles or electromagnetic waves. Down in the galaxy nucleus is the energy source which either emits energy in two beams pointing in opposite directions or, alternatively, emits energy in every direction which is then channelled by gas or the magnetic field in the nucleus into the two narrow beams whose handiwork we see in the radio band.

In some examples of radio galaxies that we can see, we find that there is a very tiny radio source centred on the nucleus of the galaxy, only a few light-years across, and that this too has a double structure along the main source's axis. This tiny source seems to be close to where the twin beam responsible for the huge double radio lobes is formed.

The enigmatic quasars

Not all the bright sources in the radio sky are associated with obvious visible galaxies. In the early 1960s the interferometer radio telescopes began to pin down the positions of radio sources in the sky with reasonable accuracy, so that it became worth while to search hard on photographic plates for any visible counterparts of the radio sources. As a result a new class of object was discovered which has remained controversial in astronomy ever since.

Let us concentrate on a particular bright radio source, '3C273', the somewhat anonymous name of the 273rd source in the third Cambridge survey of radio sources. When we search in the visible band at the position of the radio source we find, to our surprise, not a galaxy but a blue star. It is natural that at first astronomers imagined they had found a new kind of radio star in the Galaxy. We travel towards it expecting a short hop across the Milky Way, but it is still remote as we leave the Galaxy. We notice that it is in the general direction of the Virgo cluster of galaxies: perhaps it is some strange member of the cluster. We travel on to the Virgo cluster but 3C273 is still as far off as ever. Only when we have travelled over two thousand million light-years, almost as far as the most distant visible galaxies, do we arrive at the source. A galaxy at that distance would be a very faint visible object, a tiny smudge on a photograph with a giant telescope. 3C273 on the other hand is a moderately bright 'star', visible to quite a small telescope. In fact it turns out that it has been photographed many times in the last century as part of other astronomical programmes. These photographs show that this 'star' is variable in visible light: 3C273 can vary its light output by as much as 50 per cent up or down over a period of about 10 years. The brightness of the 3C273 'star' coupled with its huge distance means that it is a hundred times more powerful in visible light than the most luminous galaxies.

Hundreds of other examples of 'quasi-stellar' radio sources, or 'quasars' for short, have been found, some of them even more luminous than 3C273. The fact that they are star-like shows that they are smaller than a typical galaxy, since galaxies at similar distances would appear fuzzy and reasonably easy to distinguish from stars. An even stronger limit on the size of quasars follows from the variations in their light output. A source cannot normally appreciably change its whole light output in a time shorter than the time it takes light to cross the source. So variations on a time-scale of a year mean that the source is probably smaller than a light-year across. Some quasars have been seen to vary their visible light output in months or even days. This variability means that some quasars found their way into variable-star catalogues before their true extragalactic nature was realized. Such stars are listed by a letter code and the name of the constellation they are in. The most famous quasar listed as a variable star in this way is BL Lacertae, in the constellation of the Lizard. This has been seen to vary its visible-light output by 50 per cent in a few hours. The enormous luminosity of the quasars in visible light therefore comes from a region no larger than the solar system.

The light of ten million million suns from a region the size of the solar system? This has been very hard for astronomers to swallow. How do we know 3C273 is at this huge distance? If we spread the wavelengths of the visible band out with a prism, we notice several bright lines across the spectrum. The Dutch–American astronomer, Maarten Schmidt, who first looked at such a spectrum in 1963, could not make head or tail of these lines to start with, since he was expecting to find the spectrum of a Galactic star. Then he noticed that they corresponded to the pattern of the characteristic wavelengths of a hot gas of ionized hydrogen, but all shifted in wavelength by 14 per cent towards the red end of the spectrum. This was obviously not a star in our Galaxy as everyone had assumed at first. After considering some of the other possibilities, for example a gravitational redshift from a very condensed star, Maarten Schmidt concluded that 3C273 was a very distant object whose wavelengths were being red-shifted by the expansion of the universe. According to the Hubble law (velocity proportional to distance) which we met in the last chapter, the implied velocity of recession of 14 per cent of the speed of light, or 42 000 kilometres per second, implies a distance of about 2½ thousand million light-years.

Although 14 per cent would be a large redshift for a galaxy, this is one of the smallest redshifts seen in a quasar. We can now find quasars whose light has had its wavelengths redshifted by more than 300 per cent and which are therefore receding from us at speeds close to that of light. The velocity–distance law puts these quasars at the edge of the known universe.

Can we be sure that quasars really are at these huge distances, for it is their huge distances that force us to believe the huge light outputs? For a long while this has been a great controversy among astronomers and some still have doubts. I have to admit that I have vacillated on this question. First I believed that they really are distant. Then I became convinced for a while that some of the quasars, the ones that vary their light output for example, are much nearer to us and therefore less dramatic in their light output. The two facts that finally convinced me, and most other astronomers, that quasars really are so distant are that many of the quasars with smaller redshift, including 3C273, are found in little groups of galaxies with the same redshift, and that for some of the quasars which do not have so dazzling a luminosity in visible light we can see a fuzzy object surrounding the quasi-stellar core. This fuzz can be shown to be a normal galaxy. BL Lacertae, which we mentioned above as an example of a very variable quasar, has been seen both as a very luminous point-like quasar and as a much weaker one with an underlying galaxy showing through. However just to prove the truth of Murphy's celebrated law, that if anything can go wrong it will, when we spread the visible light from BL Lacertae out into a spectrum we cannot find any bright emission lines across the spectrum, so it is not a normal, *bona fide* quasar. The evidence for the cosmological nature of quasar redshifts looks pretty good though and scientists can't always wait around for the evidence to become perfect. Astronomers did not wait until 1838, when Friedrich Bessel announced that he could see the tiny apparent motion of the star 61 Cygni on the sky due to the motion of the earth round the sun, to accept Copernicus's picture of the solar system.

Quasars are therefore violent events in the nuclei of galaxies. A source in the nucleus of the galaxy flares up to an enormous luminosity in visible light and in most cases this makes it impossible to see the faint haze of surrounding starlight. We are dazzled by the quasar outburst. It is somewhat analagous to what happens when a supernova goes off in a near-by galaxy. For a few days the supernova outshines the whole of the rest of the galaxy. With quasars the star-like core outshines the galaxy for as much as a million years.

To give us some further anxiety about our 'cosmological' interpretation of the redshifts of quasars has come the discovery that the very tiny variable radio sources seen in a few of them appear to be expanding outwards faster than light. The theory of relativity prohibits speeds faster than light, but motion almost at the speed of light can sometimes appear to look even faster from certain angles. We have to assume that is what is happening in these 'faster-than-light' quasars.

If we now return to the radio-galaxies like M87 and Cygnus A and look more closely at the visible light from their nucleus, we can just about see faint star-like sources in their nuclei too. The only difference from a quasar seems to be in the power of this compact visible core, which is only a few per cent of the total starlight for these radio-galaxies.

M87 gives us a clue to what is driving the activity in radio galaxies and quasars. Since it is relatively near by we can study the light distribution and star motions in its nucleus. As I mentioned in the last chapter, these star motions seem to point to a dark, tiny concentration of matter a thousand million times as heavy as the sun, hidden in the very nucleus of the galaxy. This could be the first evidence of gigantic black holes in the universe, where matter can disappear as if into a celestial whirlpool. We shall get a better look at activity in galaxy nuclei and at the black-hole phenomenon on our X-ray voyage. With radio eyes we do not see very close to the mysterious source of energy in active galaxy nuclei. But radio gave us the first inkling of what is happening and therefore occupies a special place of honour in the new astronomy.

As well as discovering radio galaxies and quasars, which we will return to on our ultraviolet and X-ray voyages, the radio band demonstrated something very significant about the cosmic landscape. As we travel out to greater distances, and back in time, the proportion of galaxies that are exceptionally luminous radio galaxies and quasars increases rapdily. Our own times seem to be the calm after the great period of galaxy activity. Matter in the universe has undergone strong evolution in its form.

This was one of the fatal bits of evidence for an attractive model of the universe called the 'steady-state' theory. Put forward in 1948 by three British theoreticians, Herman Bondi, Tommy Gold, and Fred Hoyle, the theory was that the universe has always looked the same and always will. The universe expands, but new matter is created spontaneously and new galaxies form out of it, to keep the average space between galaxies and the average age of galaxies always looking the same.

How can the evolution of galaxy activity over the cosmological time-scale be explained? We already know that galaxies have changed very significantly since their birth as clouds of gas. Whether the evolution of activity in galaxies can be fitted into this picture of galaxy evolution, or implies some really new factor in the evolution of matter, remains to be seen.

4
THIRD VOYAGE
The ultraviolet landscape

The ultraviolet and earth's atmosphere

We now move to a band even further from our own immediate experience than radio, the ultraviolet. Our bodies are aware of ultraviolet radiation for it is this that gives us a sun-tan when we lie on the beach in the summer sun. We can deduce that the earth's atmosphere is stopping much of this radiation, since even at the modest altitude of a ski resort people become suntanned much more quickly that at sea level. A sheet of glass is even more effective than the earth's atmosphere at stopping ultraviolet light: one does not get suntanned indoors, even though the visible light and infrared heat still stream into the room.

In 1801 the German electrochemist Johann Ritter demonstrated the existence of ultraviolet radiation from the sun by spreading out the sun's visible wavelengths with a prism and then holding a paper soaked with silver chloride in the blank region of the spectrum beyond the violet. The plate became blackened by the 'ultra' violet radiation. However few of the ultraviolet photons from the cosmic landscape reach the ground. The ultraviolet band stretches from wavelengths of 100 ångstroms (1 Å is a hundred-millionth of a centimetre) up to the wavelength of violet light, 4000 ångstroms (0.4 micrometre), but only light with wavelengths between 3000 and 4000 Å reaches the ground from the cosmic landscape. Photons with wavelengths less than 3000 Å are absorbed by the earth's atmosphere, which is just as well for life on earth, for the shorter ultraviolet wavelengths are hostile to life. Wavelengths between 2000 and 3000 Å are absorbed by ozone in the upper levels of the atmosphere and any big reduction in the amount of ozone would let this harmful ultraviolet radiation through. The result of such an increase in radiation would be a world-wide increase in the incidence of skin cancer. Sunburn and 'snow-blindness', both caused by ultraviolet radiation, would also become much more common. Ultraviolet wavelengths are highly toxic to living cells, but they do not have very great penetrating power, so they damage only the skin and eyes.

The longest ultraviolet wavelengths, 3000–4000 Å, are accessible to ground-based astronomy and have been studied for almost a century. Because normal glass is opaque to ultraviolet light we have to use a telescope with mirrors (a reflecting telescope) rather than one with lenses (a refracting telescope). Most modern telescopes are reflectors, in fact. To detect the radiation we have to use special photographic emulsion sensitive to ultraviolet light or else some kind of electronic tube which can respond to very faint levels of ultraviolet light falling on it. We would normally use these detectors in conjunction with a filter which only lets ultraviolet light through. Alternatively we could spread the light out into its constituent wavelengths using a prism and then just study the ultraviolet part of the resulting spectrum. The prism would have to be made of special glass, for example fused silica, that is transparent to ultraviolet light.

The shorter wavelengths have had to wait until the space age before we could get a glimpse of their cosmic landscape. In 1946, Richard Tousey and a co-worker at the US Naval Research Laboratory used a captured German V2 rocket to look at the sun at ultraviolet wavelengths shorter than 3000 Å. In the 1950s and 1960s a number of rockets were launched to look at the ultraviolet sky. Recent rockets have gyro stabilizers which keep the ultraviolet detector pointing at a particular point on the sky. They also have 'star-trackers' which allow the ultraviolet telescope to be oriented with respect to chosen bright stars. These devices have allowed astronomers to study the sun, stars, and galaxies and quasars in the ultraviolet. An unstabilized rocket pitches and rolls in an uncontrolled, but trackable, way. In the process a detector on board sees many different directions on the sky and this was taken advantage of by the Naval Research Laboratory group in 1955 to make a crude survey of the ultraviolet sky.

The real breakthrough in short-wavelength ultraviolet astronomy came with satellites, starting with the first of the Orbiting Solar Observatory (*OSO*) series in 1962, followed by the at first ill-fated Orbiting Astronomical Observatory (*OAO*) series in 1972. *OAO-1*, launched in 1966, had a power supply failure before it had even looked at the sky, and *OAO-3* failed to reach its correct orbit and had to be destroyed. The expensive failures in these early efforts to use space technology for astronomy (over 400 satellites had already been launched by 1966 when *OAO-1* was launched) were a serious setback for space astronomy. When you read that a scientific satellite has failed to get into orbit, or has broken down before starting work, you should spare a thought for the hundreds of people who have spent up to five years designing and building the satellite. The failures of *OAO-1* and *OAO-3* were compensated for by the success of *OAO-2*, launched in 1968, which carried two ultraviolet experiments with seven telescopes in all. A second version of the *OAO-3* satellite was launched in 1972 and became known as the *Copernicus* satellite, in honour of the 500th anniversary of the birth of the great Polish astronomer who first demonstrated the earth's motion round the sun. The *Copernicus* satellite has had a tremendous impact on ultraviolet astronomy. We are now also beginning to get the results from the *International Ultraviolet Explorer*, launched on 26 January 1978. Ultraviolet experiments were also carried on some of the *Apollo* manned missions to the moon.

For satellite work the ultraviolet radiation is detected electronically by a grid of tiny detectors which are scanned regularly, rather like the tube of a television camera, and their signals transmitted to earth. Alternatively, in an 'ultraviolet-to-visible image converter', the ultraviolet photons impinge on a fluorescent screen which emits visible light and is viewed by a TV camera. For wavelengths shorter than 1050 Å the equipment cannot include any windows or lenses since there is no material transparent to these wavelengths. And for wavelengths shorter than 500 Å, we cannot focus the light by reflecting it directly off a parabolic metal surface — the photons just plough into the metal. Instead the light is deflected at a 'grazing incidence' off a cone-shaped mirror.

For sources outside the solar system a new barrier to ultraviolet astronomy appears at a wavelength of 912 Å. Photons of shorter wavelength than this have enough energy to knock the electron off a hydrogen atom in its neutral atomic state, to produce ionized hydrogen. In the process the energy of the photon is used up and the photon is therefore absorbed. Now interstellar space is filled with atomic hydrogen and these very-short-wavelength ultraviolet photons do not have to travel very far before they are absorbed. In fact we can hardly see outside the solar system at these wavelengths. Even the light from the nearest stars gets heavily absorbed before it reaches earth. There are, however, two factors which make the situation less hopeless than it might

71

seem. First, as we go to shorter wavelengths the fog begins to lift. In fact for every factor of ten we go shorter in wavelength the transparency improves by a factor of about three hundred. And secondly the atomic hydrogen is not spread smoothly and uniformly through the Galaxy, but is concentrated in millions of clouds. In some directions there is a bit of a gap in the clouds and we see the odd ray of light through this fog of the far ultraviolet, particularly at the short-wavelength end at 100 Å to 400 Å. A few very bright stars, which happen to be in the direction of these gaps, can then be seen at these wavelengths. It is fortunate in this respect that the sun seems to be in a relatively low density part of the interstellar gas in between the clouds. We should eventually be able to explore most of the space out to several hundred light-years from the sun. Because there is very little atomic hydrogen between us and the sun, we can see the sun quite clearly if we get above earth's atmosphere by rocket or satellite and, as we shall see, it is a fascinating sight.

What do we expect to see in our band? We remember from the visible band the colour scale for hot bodies, that as we go from red to blue we are looking at a scale of increasing temperature. In the ultraviolet, at wavelengths even shorter than blue light, we therefore expect to see a landscape of very hot gas, at temperatures from ten thousand degrees at the long wavelength end of the band to a million degrees at the shortest wavelengths.

The sun and its corona

As we set off on our voyage, we do not expect the sun to look very interesting. The temperature of the surface of the sun in visible light, the 'photosphere' is 5600 °C. The visible light from the sun is at its most intense at yellow wavelengths and we would have expected very little radiation to be coming in the ultraviolet. In the first chapter we mentioned that the light from all hot bodies has a characteristic spread of wavelengths, being strongest at wavelengths such that the photons have the same energy as the atoms of the hot body, getting weaker very rapidly as we go towards shorter wavelengths and less rapidly as we go towards longer wavelengths. As we start to scan across our band from the long wavelength, violet end, this does indeed appear to be true. The brightness of the solar disc gets weaker very rapidly as we go towards shorter wavelengths just as would be expected for a hot body at 5600 °C. However at wavelengtns of about 1500 Å and shorter the sun starts to appear much brighter than we would have expected and also somewhat larger than in the visible. This extra ultraviolet radiation is coming from gas above the visible surface.

Now, we have already seen on our radio voyage that the sun does not really have the sharp boundary that we seem to see in visible light. There continues to be low-density gas all the way to the earth and beyond, so that in a certain sense the earth is still inside the sun. How hot would we have expected this gas beyond the sun's visible surface to be? Well we can deduce that inside this surface the temperature falls off steadily with increasing distance from the centre of the sun, from about ten million degrees at the centre of the sun to 5600 degrees at the visible surface. This is because the weight of the outer layers of the sun press down on the gas nearer the centre. The gas is compressed by this pressure and the compression heats the gas up. There is a similar effect on the interior of the earth. As you go down a deep mine, it gets almost unbearably hot. At a depth of only 60 kilometres inside the earth the pressure and temperature are so high that rocks are melted, and as a result most of the earth's interior is molten.

This gradient of temperature also determines the direction of heat flow, i.e. from the centre outwards. In the centre of the sun, the heat is generated by thermonuclear reactions, the fusing of hydrogen into helium. In the centre of the earth it is the radio-activity of the rock which provides the heat flow. The structure of a star or planet is therefore a complicated balance of gravity and pressure forces on the one hand and of energy generation and heat flow on the other.

All this suggests that the gas above the visible surface of the sun should get colder with increasing distance from the sun. The enhanced ultraviolet radiation that we are seeing from above the visible surface shows that this is not so. For the first few hundred kilometres above the visible surface, the temperature does indeed fall off with distance as expected, reaching a minimum value of 4000 °C. Beyond this minimum, however, the temperature begins to increase as we go outwards, slowly at first and then more rapidly. It gets to 20 000 degrees in the first few thousand kilometres, then undergoes a rapid transition from 20 000 to 500 000 degrees in the next few hundred kilometres. Above this sharp transition region the temperature continues to increase but at a much slower rate, finally reaching a temperature of one and a half million degrees at about 10 000 kilometres above the sun's visible surface. We are now about 1 per cent of the sun's visible radius above the surface of the sun. This million-and-a-half-degree gas extends out to distances several times the sun's radius before eventually merging with the interplanetary medium.

This region of very hot gas around the sun is called the 'corona' and has probably been known to man for thousands of years, since the inner part of the corona can be seen as an irregular halo round the sun during a total eclipse. The main excitement of a total eclipse, apart from the drama of the sudden night, is the brief glimpse we get of the corona, which never looks quite the same on any two occasions. We are seeing the hot coronal gas only indirectly in fact through its scattering of visible light from the solar disc. In the past astronomers made elaborate expeditions to observe solar eclipses. Any particular eclipse is visible from only a small fraction of the earth's surface and although total eclipses are quite common, they are not often visible at convenient locations on earth. Nowadays if astronomers want to study the corona they make their own eclipses by blocking out the sun's disc with a small circular piece of metal.

It was in 1933 that it was first discovered that the corona is at a temperature of a million degrees. This seemed paradoxical at first. How could the sun, at a temperature of 5600 degrees, heat this gas up to a million degrees? This seemed to violate the second law of thermodynamics, that heat cannot flow continuously from a colder body to a hotter one. The explanation is that the heating of the corona is nothing to do with the radiation from the visible surface of the sun. Instead the bubbling, boiling motions of the convective zone below the sun's surface generate sound waves. These sound waves travel out from the sun's surface and are then dissipated by friction in the gas of the corona. The energy of the sound waves is turned into heat by the friction and the corona is therefore heated to a high temperature. It's analogous to what happens when a human being, with a body temperature of only 37 °C, rubs two sticks together to generate a temperature of several hundred degrees to light a fire.

The real fascination of our ultraviolet picture of the sun comes when we split the radiation up into its constituent wavelengths — using a prism, or 'grating' spectrograph, to spread the wavelengths out into a spectrum of wavelengths. A 'grating' is a piece of glass with very fine parallel lines etched on it, tens of thousands to the centimetre. It has a very similar effect to a prism in deflecting light of different wavelengths by different amounts, with the result that the different wavelengths are spread out into a spectrum. We pass the light from the sun through a narrow slit into the spectrograph. If one particular wavelength is very prominent, it appears as a bright line across the spectrum.

Across the ultraviolet spectrum of the sun we see thousands of lines, characteristic wavelengths of all the more abundant elements: hydrogen, helium, oxygen, nitrogen, carbon, silicon, magnesium, neon, and iron. Let us now consider how some of these characteristic wavelengths arise. In general they are emitted when an electron jumps from one of its permitted orbits to another, releasing a precise amount of energy in the form of a photon with a precise wavelength or frequency.

Let's look at the brightest line across the sun's ultraviolet spectrum, the Lyman α-line of hydrogen at 1216 Å. This is the most important wavelength of a series of wavelengths characteristic of hydrogen, called the Lyman series of wavelengths after their discoverer. The other wavelengths fall between 912 and 1216 Å. The Lyman series of wavelengths arise from hydrogen that is ionized, i.e. in the form of positively charged nuclei (protons in the case of hydrogen) or 'ions', and electrons moving around freely. Every so often, in their helter-skelter motion through the gas a proton and electron meet and combine together to form a neutral atom. Each time this happens the electron in the hydrogen atom has a certain probability of finding itself in an excited, high energy state. It cannot remain in this state for long and the electron soon drops to a less excited state, emitting a photon. It is some of these photons which make up the Lyman series. When the electron drops from the second lowest energy state to the lowest energy state or 'ground state' Lyman α-photons are emitted. At least six lines of the Lyman series can be clearly seen across the sun's ultraviolet spectrum.

The neutral atom does not remain in that state for long if the gas is hot, or if it is being illuminated by a strong source of far-ultraviolet radiation. Collisions with other atoms, or alternatively the absorption of a high energy photon, soon knock the electron off the atom again and restore the gas to its ionized condition. The photon would have to have a wavelength shorter than 912 Å in order to ionize a hydrogen atom. For this reason far ultraviolet radiation and the even higher energy X- and gamma-radiation which are the subject of our next voyage are often called 'ionizing' radiations.

The other abundant atoms also have characteristic wavelengths in the ultraviolet. Things get more complicated the heavier the atom is, since a heavy atom can have one, two, three, or more electrons stripped off. Each of these states has its own series of lines. Which wavelength band they occur in depends on the energy needed to strip the relevant number of electrons off. For example, iron with all 26 electrons stripped off has its lines in the X-ray range.

The different atoms and their different states require different densities and temperatures to produce strong spectral lines. For example the characteristic ultraviolet wavelengths of ionized oxygen with five electrons stripped off (denoted O VI: the Roman numeral corresponds to one more than the number of missing electrons — O I means neutral atomic oxygen) requires a temperature of about 100 000 degrees, while magnesium with nine electrons stripped off (Mg X) needs 1.4 million degrees. By looking in the light of these particular ultraviolet wavelengths we see only the gas at these particular temperatures. In this way we build up our picture of how the temperature varies across the sun's corona.

We soon see that the corona does not have a homogeneous structure. The first two thousand kilometres above the visible surface, called the 'chromosphere' because of its red colour due to a bright characteristic wavelength of hydrogen at 6563 Å in the red, do form a fairly uniform layer. Above this, however, many small jets or spikes of relatively cool gas, called 'spicules', shoot up into the corona. They are typically 1000 km thick and about 5000 km tall and they would look like a forest of bare trees standing up on the surface of the sun, but for the fact that they are transient. They rise and fall continuously, each one lasting about five minutes.

Another transient phenomenon which we can study in the ultraviolet is that of active or eruptive prominences, which appear over a sunspot group. A stream of hot gas, at a temperature of 30 000 degrees, surges up to 400 000 kilometres above the surface of the sun before falling back to make a great arch or loop. Round an active region or sunspot we see a continuous surging of loops and streamers rising and falling in a chaotic manner. Even more impressive is the solar flare, during which the sun brightens up at ultraviolet wavelengths by a large factor. We have already seen that a solar flare gives rise to bursts of radio radiation. A solar flare will appear more dramatic than in radio or ultraviolet, though, in the X-ray band.

The corona does not present a steady appearance to us. It varies in brightness over the 11-year cycle of solar activity. Near the minimum of the solar cycle, when almost no sunspots are seen, the corona has a temperature of 1.2 million degrees. Near maximum activity the corona is nearly twice as dense as at minimum and is hotter, at a temperature of 1.8 million degrees. The hotter regions over active sunspots give the corona its irregular appearance.

We have spent some time on the ultraviolet appearance of the sun since this band has given us so many surprises about our own familiar star. Also for a major part of our band (the wavelengths between 400 and 912 Å) the sun is the only astronomical object we can study in any detail so far because of the fog of interstellar atomic hydrogen.

Earth, Venus, and the zodiac

We look back at the earth. To our surprise the earth is cloaked in a corona of its own, the 'geocorona'. The lightest gas in the earth's atmosphere, hydrogen, extends out far beyond the familiar and denser gases we breathe, out to 100 000 km from the earth in fact. This hydrogen halo round earth absorbs ionizing photons from the sun, is ionized and then recombines to give the characteristic wavelengths of the Lyman series, particularly Lyman α. The result is the ultraviolet geocorona.

We travel now to the planet Venus, which in the visible band revealed nothing more to us than the featureless yellow tops of its clouds. We now see a complex pattern of bright and dark swirls crossing the whole surface of the planet. These are presumed to be due to an uneven distribution of some atmospheric constituent that absorbs ultraviolet light, but astronomers don't know yet what the absorbing material is. If we watch the surface of the planet for some hours we will notice that some of the features have moved, suggesting high wind speeds in the upper atmosphere, up to 100 metres per second. On earth such wind speeds are only encountered in narrow jet streams.

We start to leave the solar system. All along the plane of the solar system we see the faint glow of the zodiacal light, the sun's ultraviolet light scattered by small grains of interplanetary dust spread in a cloud through the ecliptic plane. As we look round the ecliptic plane towards the sun the zodiacal light seems to merge into the outer portion of the corona. This outer portion of the corona (at a distance more than twice the solar radius from the centre of the sun) is in fact nothing to do with the coronal gas, but is due to the deflection of light from the solar disc by dust particles near to our line of sight to the sun. As we reach the boundaries of the solar system we see a glow at the wavelengths of 1216 Å, the Lyman α-line of hydrogen, and 584 Å, the corresponding line of helium. They are seen because the light of the sun at these wavelengths is reflected back off hot, low-density gas between the sun and the nearest stars. This gas is streaming past the solar system at a speed of 20 kilometres per second. It is possible that the sun may be passing through the edge of an interstellar cloud only a tenth of a light-year away.

Hot stars, coronae, and planetary nebulae

We leave the solar sytem behind. The absorbing effect of the interstellar atomic hydrogen now confines us to wavelengths greater than 912 Å or, for a few near-by stars, to wavelengths shorter than 400 Å where the fog starts to lift. We see a landscape of hot stars, with temperatures in the range 10 000 to 100 000 degrees. The stars looked blue or white in the visible band, but they are really 'ultraviolet' stars for their main energy output is in this band. When we spread their light out into its constituent wavelengths, we see across their spectra the characteristic wavelengths of hydrogen and helium. Most of the stars are very massive and luminous, as heavy as ten to fifty of our suns and more luminous than a hundred to a hundred thousand suns. However we also see the diminuitive and feeble, but very hot white dwarfs, cooling off to their final cold dark state. The white dwarf, HZ 43, 200 light-years away in the constellation of Coma Berenices, was the first source outside the solar system to be detected in the extreme ultraviolet wavelength range, 100 to 300 Å.

Round the hot luminous giant stars we see an immense cloud of hydrogen lit up by ultraviolet radiation. The ionizing photons (wavelength less than 912 Å) ionize this cloud out to a distance which depends on the density of the gas and the number of ionizing photons being emitted per second by the star. The size of the cloud of ionized hydrogen, called an H II ('H-two') region, can range up to hundreds of light-years. These hot giant stars therefore affect their environment on a far larger scale than the sun.

When we look more closely at some of these hot luminous stars we find a very violent wind of gas blowing off their surface at speeds greather than a thousand kilometres per second. We can tell this by looking at the way certain characteristic wavelengths of ionized silicon and carbon, with three electrons stripped off, are spread out by the Doppler shift due to the motion of the hot gas. These stars, so luminous that they are called 'supergiants', blow off a mass equal to that of the sun in only a hundred thousand years. This is a rate of mass-loss a million times more rapid than that of the sun's wind. These winds create huge bubbles of hot gas around the blowing supergiant stars. This could provide the explanation for another surprise of the ultra-violet cosmic landscape. We saw in the radio band that the space between the stars is filled with clouds of cold atomic hydrogen. The hot stars which are strong ultraviolet emitters also create those huge zones of ionized hydrogen around them at ten thousand degrees, the H II regions. If we survey the Milky Way at an ultraviolet wavelength characteristic of ionized oxygen (O VI, with five of oxygen's six outer electrons stripped off) which is excited only at very high temperatures, we see that interstellar space is pervaded with thin, low-density gas at a temperature of between a hundred thousand and a million degrees. This sizzling hot gas does not make up much of the total mass of gas between the stars, but how it got to this temperature is quite a puzzle. The bubbles blown out by hot supergiant stars, or the shells of hot gas created by supernovae explosions, eventually merging together when they get large enough, are possible explanations.

Although most of the red and yellow stars that are so common in the visible sky have almost vanished from sight in the ultra-violet landscape, a few surprise us by being much brighter than expected in the ultraviolet. The reason is similar to that for the sun, in that they are surrounded by coronae of hot gas. Many near-by stars like the sun can be inferred to have coronae from ultraviolet lines in their spectra. So too can some luminous stars at the red giant stage of their evolution and also some much younger red objects still in the process of becoming stars, the T Tauri stars, named after a prominent star of this type in the constellation of Taurus. We will return to T Tauri stars on our infrared voyage.

Another fascinating type of object can be seen by travelling towards the constellation of Lyra, to the Ring nebula. We see what looks like a circular ring of hot ionized gas, looking rather like a smoke ring, surrounding a very hot star, even hotter than the giants and supergiants that cause H II regions. This is an example of a 'planetary nebula', a spherical puff of gas blown off a dying star in its last convulsion before cooling down to become a white dwarf. Although the hollow shell of gas is thrown off in all directions from the star we only see a bright rim round the edge of the shell, where we are looking through more gas. The temperature of the star is around a hundred thousand degrees so the bulk of its radiation must come out in the very highest energy, shortest wavelength photons of our band, around 300 Å.

The puff of gas that the star has thrown off contains about a tenth of the mass of the parent star, which probably lies in the range one to four times the sun's mass. The hot star is the naked core of the star, left after it has blown off its huge red giant atmosphere. About 700 planetary nebulae have been discovered all around the Milky Way. It seems that most stars with masses between that of the sun and about four times larger become planetary nebulae in the final stages of their evolution after their time as red giants. So, as we look at the Ring nebula we can imagine that we are seeing the last days of the sun, thousands of millions of years hence. Have the inhabitants of earth managed to remove themselves to a planet near a younger star or did they watch helpless as the sun grew to be a red giant and engulfed them?

Clouds of gas and dust

As we look round the Milky Way in the ultraviolet waveband, we see that the effects of dark clouds in cutting out the light from distant stars is even more severe than it was in the visible. One curious feature is that at a wavelength of about 2200 Å in the ultraviolet, the absorption of starlight by this intervening material is very much worse than at other ultraviolet wavelengths. This seems to require its own special explanation, very tiny grains of graphite spread through interstellar space absorbing the radiation.

Sometimes we can see a hot star radiating strongly in the ultraviolet behind a cloud of cold gas that is not too dense. This allows us to see whether the ultraviolet photons of the star are absorbed by the gas in the cloud. We expect the different atoms in the cloud to leave their imprint on the light from the star by removing their characteristic wavelengths, leaving dark lines across the spectrum. A good example is the star ζ-Ophiuci, in the constellation of the Serpent Bearer. The gas cloud in between this star and earth has a density of about one thousand hydrogen atoms per cubic centimetre, quite high for an interstellar cloud, and a temperature of about −160 °C. The most common elements do indeed absorb out their characteristic wavelengths from the star's light but not as strongly as we might have expected for a cloud of this density. Either all the elements except hydrogen and helium have a much lower abundance in this cloud than they do elsewhere in the Milky Way or, more likely, they are not in the form of atomic gas. The most likely solution is that 90 per cent or more of the 'heavy' elements are in the form of dust grains, whose existence was strongly suggested in the visible band and which become of overwhelming importance in the infrared band, as we shall see later.

One of the most important determinations of element abundance that can be made in the ultraviolet is that of heavy hydrogen or deuterium. An atom of deuterium is identical to a hydrogen atom except that the nucleus has a neutron in it, in addition to the proton which makes up hydrogen's nucleus. Now deuterium is quite easily made in the centre of stars, but it is also destroyed just as easily because the proton and neutron in the nucleus are not very tightly bound together. From processes in stars alone we should expect there to be no deuterium at all in the interstellar gas, since by the time a star gets round to blowing gas off into space, any deuterium would all have been destroyed. To make deuterium we need a hot oven, but when the deuterium is cooked it has to be taken out of the oven very quickly, more quickly even than happens in stellar explosions.

When we look towards a hot bright star like Spica, the brightest star in the constellation of Virgo, to test the abundance of deuterium from its Lyman series spectral lines, we find a small but very significant abundance, 0.002 per cent of that of hydrogen. The only explanation that anyone has managed to come up with for this deuterium is that it was produced in the fireball phase of a hot 'Big-Bang' universe. We shall be exploring this stage of our past on our microwave voyage. The crucial aspect of the Big Bang for deuterium is that the universe is cooling down so fast that some of the deuterium which formed when the temperature was high manages to survive. The fireball is the hot oven which cools off at the rate needed for deuterium to survive. Thereafter deuterium gets destroyed if it finds itself inside the hot core of a star, but otherwise survives. This tiny abundance of deuterium is one of the most important pieces of evidence that we live in a Big Bang universe.

Galaxies and their active nuclei

We are ready to leave the Milky Way and start to explore the landscape of galaxies. In the Andromeda galaxy and other nearby spiral galaxies we see in our band their hot young stars, tracing out the spiral arms where they were formed. Most of the elliptical galaxies, with their older red stars, are much harder to see except in the longer wavelengths. However if we travel to the giant elliptical M87 in the Virgo cluster of galaxies and look at its nucleus, we notice something interesting. First we see the characteristic wavelengths of a massive cloud of ten-thousand-degree gas. Then from a very compact, point-like core in the heart of the nucleus we see ultraviolet radiation of all wavelengths. This does not have the typical distribution of wavelengths of hot gas, but seems to have some more violent origin: for example, radiation from electrons moving very close to the speed of light. This compact core is not nearly so noticeable in the visible band because the radiation from the stars swamps it. It is the ultraviolet emitting core which is exciting the gas surrounding it. We are reminded of the other evidence for activity in the nucleus of M87, the very strong radio emission and the rapid motions of stars in the nucleus indicating some dark massive object.

We start to look at the nuclei of galaxies in our neighbourhood more closely, not just at the radio galaxies which we already know to be active. Soon we come across an entirely new kind of active galaxy, the prototype of which is an apparently unremarkable galaxy NGC 4151, number 4151 in the New General

Catalogue of galaxies. This catalogue is the modern revision of the lists of nebulae published by William Herschel and his son John between 1786 and 1864. NGC 4151 is a spiral galaxy about 50 million light-years from earth. When we look carefully at ultraviolet wavelengths we see that in the midst of the starlight from the disc and nucleus of the galaxy there is, as in M87, a tiny pointlike source of non-thermal radiation. The visible and ultraviolet spectrum of this galaxy shows, also like M87, the characteristic wavelengths of hot gas which is being illuminated by a non-thermal core. The bright lines on the spectrum corresponding to these wavelengths look exceptionally broad, the characteristic wavelengths of hot gas which is being illuminated by the tens of thousands of kilometres per second. This property was noticed in the visible band by the US astronomer Carl Seyfert in the 1940s and he made a list of several galaxies with these very rapid motions; these have become known as Seyfert galaxies. We can find these galaxies, either from their spectra, or from their strong ultraviolet emission. The Armenian astronomer, B. E. Markarian, has been compiling lists of galaxies which have excess ultraviolet emission and about one-tenth of these are Seyfert galaxies. About 2 per cent of all galaxies are Seyferts, so they are quite common in the cosmic landscape.

Another thing we notice about NGC 4151 which is very suggestive of violent activity in the nucleus of the galaxy is that when we look in the light of one of the hydrogen lines (we could use Lyman α, but this effect was discovered with a line in the red part of the visible band), we see dozens of irregular filaments of gas sticking out from the centre of the galaxy in all directions. They make NGC 4151 look like a gigantic Crab nebula. Are the nuclei of Seyfert galaxies therefore undergoing something like a supernova explosion but on a much larger scale? Since so many galaxies are in this state, it is hard to believe that we are seeing a simple explosion. More likely we are seeing some more continuous and regular activity, a phase which perhaps all galaxies go through at some time or from time to time. A small minority of galaxies show great peculiarity in their appearance, with strange plumes and arms sticking out at odd angles. Many of these seem to be caused by gravitational interaction with other near-by galaxies, rather than by activity in the galaxy nuclei, so they may not be connected with Seyferts and other active galaxies.

We are struck by the similarity of these cores in the nuclei of

Seyfert galaxies to the quasars that we found in the radio band. The nucleus of NGC 4151 is rather like a miniature quasar in its visible appearance, a point-like source of non-thermal radiation. And sure enough when we travel in our ultraviolet band to the quasar 3C273 we find that the non-thermal radiation that dominates the visible light extends out to the shortest ultraviolet wavelengths we can observe (the 912-Å limit imposed by the interstellar hydrogen). As well as the continous emission spread out over all the ultraviolet wavelengths, we notice that the ultraviolet spectra of both NGC 4151 and 3C273 are dominated by two very bright lines across the spectrum, the characteristic wavelengths of the Lyman α-line of ionized hydrogen (1216 Å) and of ionized carbon with three electrons stripped off, C IV (1550 Å). Of course, because of the redshift we see these lines at slightly longer wavelengths (14 per cent longer in the case of 3C273). When we try to study from a satellite the Lyman α-radiation emitted by a fairly near-by Seyfert galaxy, not much redshifted, we have the problem that the Lyman α from the galaxy is all mixed up with the Lyman α-emission from the earth's geocorona. This happens in fact for NGC 4151, whose wavelengths are shifted by only 0.3 per cent, but the Lyman α from 3C273 is redshifted well clear of the geocoronal Lyman α. These two characteristic ultraviolet wavelengths are very important for finding the redshifts of quasars. Once the redshift is 100 per cent or more, so that wavelengths are more than doubled, they begin to be shifted into the wavelength range accessible to the ground-based astronomer.

There seem to be only two differences between 3C273 and NGC 4151. One is the enormous difference of scale. 3C273 is a thousand times stronger than the NGC 4151 source, in ultraviolet and visible radiation. Even if there were a normal galaxy of stars surrounding 3C273 it would be impossible to detect against the blaze of light from the core.

To find the second main difference we have to shift to the radio band. 3C273 is a powerful radio emitter, both from the point-like core itself and from a jet of emission stretching out far beyond the edge of any possible galaxy there. In many other quasars we find the double source with the two components separated by up to millions of light-years, typical of powerful radio galaxies. We recall, however, that these radio galaxies are always associated with elliptical galaxies, whereas NGC 4151 is a spiral galaxy. For some reason we don't understand, active ellipticals are very powerful emitters in the radio band whereas active spirals are not. The radio-emitting quasars must therefore be in elliptical galaxies to be consistent with this.

We therefore seem to have two unrelated types of active galaxy, the radio galaxies and the radio-emitting quasars, which are found only in elliptical galaxies, and the Seyfert galaxies, which are spirals. Are there any active spiral galaxies with quasar-like power, or are the nuclei of spirals never as dramatic as those of ellipticals? To test this we need to see whether all quasars are strong radio emitters. To check this we first return to the long-wavelength end of our ultraviolet band and look at all the very blue stars on the sky one by one. We first see whether they have excess ultraviolet emission compared with normal blue stars. We then examine the spectrum of the blue stars which are strong in the ultraviolet. Sometimes we find the spectrum of a star in our Galaxy. But often we find the redshifted lines of hot gas across the spectrum and we have a quasar. If we look all round the sky we would find millions of quasistellar objects in this way. The interesting thing is that only a few per cent of them are strong radio-emitters.

The radio quasars, though discovered first, are in fact not the typical quasar. It is the majority type, the radio-quiet quasar, which is the true analogy of NGC 4151's nucleus. It is my belief that these radio-quiet quasars lie in spiral galaxies and that their only difference from Seyfert nuclei is one of scale, but this is an idea that is hard to test.

The hydrogen molecule

One last journey in the ultraviolet band points the way forward towards our sixth voyage, in the microwave band. There we will be able to study interstellar molecules, with one supremely important exception. Molecular hydrogen, easily the most common cosmic molecule, does not radiate in the microwave band. It does, however, have many characteristic wavelengths in the ultraviolet region of the spectrum. When we travel back towards a bright star along the path of photons with one of these wavelengths then if any molecular hydrogen lies in our path, there is a chance that the photon will have been absorbed by the hydrogen. When we look at the ultraviolet spectrum of a near-by star like ζ-Ophiuci we do indeed find dark lines across it at these wavelengths, showing that a cloud of molecular hydrogen has absorbed most of the photons at these wavelengths. We shall return to the subject of molecular clouds later on, but we can note that the clouds we can spot in the ultraviolet cannot be very thick and dark otherwise we wouldn't be able to see the star through it.

The ultraviolet is a paradoxical band. At the shortest wavelengths we can see only the sun, but at the longer wavelengths we can study the most distant objects yet known to us, the quasars. Above all it is the band of matter at temperatures of ten thousand degrees or more.

When will the interstellar fog that came down on us at wavelengths shorter than 912 Å clear again? We have to move to an even more alien and hostile band, the X- and gamma-ray band, to find radiation penetrating enough to get through to us.

5
FOURTH VOYAGE
The X- and gamma-ray landscape

Radioactivity and the terrestrial environment

The cosmic landscape in X- and gamma rays is the harshest that we shall see. The very names given to these radiations suggest mysterious and deadly rays. To me it never ceases to be a strange and slightly chilling experience to have an X-ray taken of my teeth or chest, to see the 'skull beneath the skin'. When the German physicist Wilhelm Röntgen discovered X-rays in December 1895 one of the first X-ray photographs he took was of his own hand. Within weeks the news had spread all over the world and X-rays were being put to use. Knowing nothing about the safe dose, hundreds of early workers died of cancer and radiation sickness.

We have shifted to wavelengths a thousand million times shorter than radio waves, to frequencies a thousand million times higher. Now the amount of energy a photon carries increases in proportion to the frequency, so each photon of our band carries about a thousand million times as much energy as a radio photon. It is this fact that gives X- and gamma rays their tremendous penetrating power.

. Our band can be divided into two sub-bands. Wavelengths longer than about 0.02 Å, that is two thousand-millionths of a millimetre, are called X-rays and those shorter than this are called gamma rays. The longest wavelength X-rays, those of wavelength 10–100 Å are called 'soft' X-rays, while the shortest wavelength X-rays, 0.02–1 Å, are called 'hard' X-rays. The boundaries are rather arbitrary. As we go to the longer and less penetrating wavelengths of the soft X-rays we start to experience the fog of the interstellar gas. To penetrate the human body and make an X-ray photograph, rather hard X-rays are needed, of wavelength around 0.1 Å. More energetic photons of shorter wavelength still may be used to try to destroy cancerous cells. Gamma rays are the deadliest component of radioactivity and their effect on the human body in large doses has been gruesomely demonstrated at

Hiroshima and Nagasaki.

It is worth trying to understand radioactivity in some detail, since this is one of the ways we experience the cosmic landscape in our band. A radioactive substance like uranium, radium, or plutonium continuously emits three types of 'ray': alpha, beta, and gamma rays. Alpha rays are in fact the nuclei of helium atoms, thrown off as the radioactive atomic nucleus transmutes to a different element of lower atomic number (with fewer electrons orbiting round it and fewer protons and neutrons in the nucleus). Alpha rays are not very penetrating and can be stopped by a sheet of paper or a few centimetres of air. Most of the energy given off by a radioactive substance is in this harmless form.

Beta rays are quite a bit more penetrating and can pass through several millimetres of aluminium. They too are not really rays, but atomic particles, this time electrons. Although an electron is thousands of times lighter than a helium nucleus, these electrons are moving very fast and this is what gives them their penetrating power.

The gamma rays are the ones that do the damage. They are so penetrating that they can pass through 20 centimetres of iron or several centimetres of lead. Of the three types of radioactive 'ray', gamma rays are the only ones that are really rays. They are electromagnetic radiation, or light, just like visible or radio waves are. The effect of X- and gamma radiation on living tissue is measured by a quantity called the rem. The dose from a dental X-ray is about one-sixtieth of a rem. The human body can some-times survive a single dose of radiation as strong as 250 to 800 rems, but it is exceedingly unlikely towards the higher end of this range. Even for those who survive, the long-term effects of such a dose would be sterility, stunting of growth in children, cancer, and a general shortening of the life-span.

X- and gamma rays also increase the chances of mutations, errors in the genetic code. Most mutations are harmful, but a few are beneficial to living creatures in that their offspring are changed in a way that increases their chances of survival. The effect of high energy radiation in increasing the mutation rate is measured by what is called the 'mutation-rate-doubling dose', which is the total dose of X- or gamma radiation needed to double the mutation rate compared with the rate of mutation in a normal terrestial environment. This mutation-rate-doubling dose has been found in laboratory experiments on mice to be about 30 to

100 rems over their reproductive life. It is not known exactly what the dose is for human beings, but it is believed to be about the same as the dose needed for mice, 15–150 rems over a thirty-year period.

We tend to think that radioactivity is solely the creation of scientists and governments manufacturing and testing nuclear weapons. I am sure that most people have the impression that before the nuclear age the terrestrial environment was free of radioactivity. However this is far from the truth. The walls of our houses, the stones in our garden, the rocks on the hillside, the earth itself, all are radioactive, and always have been. The food we eat and the water we drink are radioactive and over the course of a year our own bodies subject us to a dose of about one-twentieth of a rem. A normal terrestrial environment gives us a dose of between one-fiftieth and one-sixth of a rem in a year. Certain locations on earth are exceptionally radioactive. The monozite sands of Kerala, India, subject the local inhabitants to between one-third of a rem and 3 rems per year, which is above the recommended maximum dose of one-half of a rem per year.

This X- and gamma radiation from the terrestrial environment, and the mutations they cause, have been essential for our evolution. For the most part these radiations are harmful to us yet, paradoxically, without them we would not be here. Our awareness of the cosmic landscape in this band is thus rather subtle. On the one hand, cancer. And on the other, our very existence.

How then did this radioactivity of the terrestrial environment arise? The naturally occurring radioactive substances like radium and uranium were made during supernovae explosions. They are a reminder of the violent past that the matter of earth and of our bodies has experienced. Another important cause of damage to living cells in the terrestrial environment is cosmic rays, those atomic particles moving close to the speed of light. Even the lower energy cosmic rays generally have much more energy than alpha or beta rays. Higher energy cosmic rays smash into molecules of air in the atmosphere and turn into a shower of atomic particles and photons, including gamma rays. The effect of cosmic rays on living cells is normally equivalent to about one-thirtieth of a rem per year at ground level, but it doubles for every 5000 feet you go up in altitude. A solar flare increases this dose at ground level by a factor of fifty. Between 30 and 600 kilometres above the earth's surface the normal dose from cosmic rays is about 7 rems per year and is increased by a factor of a thousand during a solar flare. Manned space missions are therefore not launched during the active phase of the sun's eleven-year cycle, when flares are quite common.

X- and gamma rays from the cosmic landscape do not penetrate to the ground, but are absorbed by the atmosphere. To get a glimpse of this dangerous landscape we have to climb above earth's protecting atmosphere. The birth of X-ray astronomy took place in 1948 when the American T. R. Burnight detected X-rays from the sun using a German V2 rocket. Fired straight up in the air, the rocket soared above the atmosphere for a few minutes. This was long enough for a photographic emulsion, wrapped in metal foil to cut out other wavelengths, to be blackened by X-rays from the sun before the rocket dropped to earth. In the following year, Herbert Friedmann and his co-workers at the US Naval Research Laboratory, demonstrated conclusively that these X-rays were coming from the sun, using two Geiger counters also in a German V2 rocket. The two main kinds of detector used to detect photons from the X- and gamma-ray cosmic landscape are 'proportional' counters — typically used for wavelengths longer than about 0.5 Å — and 'scintillation' counters, used for the shorter wavelengths. The proportional counter consists of a small chamber of gas, atoms of which become ionized when an X-ray photon passes through. The number of atoms ionized can be measured and is proportional to the energy of the X-ray photon. In the scintillation counter a crystal of sodium iodide responds to X- and gamma rays by emitting visible light. One problem with these devices is that they tend to respond to high energy charged particles too and go wild if the observing platform passes through the earth's radiation belts. Sorting out which of the 'counts' are due to X-rays and which are due to charged particles is not always easy.

At the hardest X-ray and gamma-ray frequencies the penetrating power of the photons means that we do not have to get quite so high above the atmosphere to detect them and much of the pioneering work has been done from high-altitude unmanned balloons. It was the advent of astronomical satellites, however, that opened up the cosmical landscape in the X- and gamma-ray band to our eyes. The *Uhuru* satellite, launched off the coast of Kenya in 1970 made the first all-sky survey in the

X-ray band and discovered most of the types of X-ray source that we know about today. As I write there are four other major satellites working in our band: the *Copernicus* ultraviolet and X-ray satellite, the *Ariel-5* X-ray satellite, and *HEAO-1* and *-2* new X-ray satellites with high sensitivity and very accurate pointing.

The sun and its flares

We depart on our fourth voyage. The sun has a blotchy and bloated appearance, for we are mainly seeing radiation from the million-degree gas of the corona, well above the sun's visible surface. This hot gas does not have a smooth distribution and there are huge dark holes covering much of the solar disc, where the gas is slightly cooler. We also see bright X-ray emission above active regions, sun spots, and prominences.

A solar flare is an awe-inspiring sight in our band. In the visible waveband the sun brightens by about 1 per cent during a flare, but in the X-ray band it may brighten by more than a factor of a thousand. The energy of a solar flare originates in the strong magnetic field of a sunspot, which causes a catastrophic implosion of gas just above the surface of the sun. Two streams of cosmic rays are created, one outwards to generate the radio bursts and eventually batter against the earth's magnetosphere and cause magnetic storms and aurorae, and one inwards. The inward beam of cosmic rays blasts down into the gas of the sun's surface layers, which then radiate hard X- and gamma rays in bursts that last a few minutes at a time. The sun's surface immediately below the flare is temporarily heated to ten million or a hundred million degrees and then radiates in X-rays for between ten minutes to an hour. Elsewhere in the X- and gamma-ray landscape, it will be these same two kinds of radiation that we shall be seeing, radiation generated by cosmic rays interacting with matter (or with lower frequency radiation) and radiation from gas at temperatures of a million to a hundred million degrees.

Into the stellar graveyard — white dwarfs

We leave the solar system and start to look round the Milky Way for other sources in our band. As with the early days of radio-astronomy, the first X-ray astronomers had very primitive telescopes and knew only roughly the direction of the sources they found. The brightest sources are therefore named after the constellation they are nearest, with a number. For example the brightest X-ray source in the sky apart from the sun is called Sco X-1 ('sco-ex-one') and is in the constellation of Scorpio, the Scorpion. This was the first X-ray source outside the solar system to be found and was discovered with rocket-borne Geiger counters flown by the American Science and Engineering and Massachusetts Institute of Technology groups, of Cambridge, Massachusetts, in 1962. They found this source by accident, since they were in fact trying to detect X-rays from the moon. ,

Let us travel towards this source in Scorpio. The X-rays come from a tiny region and we have to come in very close to see what is happening. We see that the X-rays come from the surface of a dark object only ten thousand kilometres across, about the size of the earth. The X-rays are being emitted by hot gas, hundreds of millions of degrees in temperature, which is falling onto this compact dark object at a speed of several thousands of kilometres per second. It is the impact of this gas thudding against the surface of the compact object which generates the enormous heat. We see that the surface gives slightly under the impact so that the compact object is not a solid object like a planet, but is gaseous, albeit extremely dense. Anyway the strong gravity we are experiencing tells us that the object is far too massive to be a planet and in fact despite its tiny size the object is about as massive as the sun.

To find out what we have here, we switch to the visible band. We find ourselves in a double-star system where the two stars are very close together, a 'close binary'. The 'primary', more massive star which we did not notice in X-rays has finished burning hydrogen and started trying to grow to become a red giant. Because of its tiny companion it does not have enough room to do this. As its radius grows to be about half the distance between the stars, the secondary's gravity starts to distort the shape of the primary into a pear shape, with the stalk of the pear pointing towards the secondary. As the primary tries to grow into a giant it becomes more and more distorted, until the protruberance towards the secondary has the form of a cone with a point on the end, the 'Roche' point. If the primary tries to grow any more, this point is like a leak and matter squirts out through it towards the secondary. If the secondary is a tiny star with very strong gravity, so the gas reaches huge speeds before it hits the star and therefore gets heated to very high temperatures, then we may see an X-ray source when this happens.

Because the two stars are orbiting round each other rapidly, the gas does not fall directly onto the compact star but forms a rotating disc of gas round it. The gas from the inside edge of the disc falls on to the compact star to give the X-rays. The gas from the disc itself, which is rather hot, radiates in the ultraviolet and visible bands. Most of the visible light we receive from the Sco X-1 system comes from this disc of gas around the compact secondary, rather than from the primary star. The companion star, the 'secondary' from which we were seeing the X-rays, is a white dwarf, a star already dead. The star has run out of nuclear fuels to burn and collapsed down to a very high density. From the earth we can see some visible light from the primary with a big telescope but we cannot see the white dwarf since its contribution to the total light of the system is negligible. We could, in principle, detect the white dwarf itself through certain lines in the spectrum of light from the whole binary system. If we look carefully at the two sets of spectral lines from some binary systems we can actually see the motions of the two stars as they orbit towards us and away from us by the different Doppler shifting of the two sets of lines backwards and forwards across the spectrum. This is assuming that the plane of the orbit of the two stars has some tilt to our line of sight, otherwise there would be no motion of the stars towards and away from us.

The speed with which the gas hits the surface of the white dwarf is a measure of the strength of gravity there. Gas from the sun would if undeflected fall onto the earth with a speed of a few kilometres per second, and this is also the speed a rocket needs to achieve if it is to escape completely from the earth's gravity, the so-called 'escape speed'. At the surface of the sun, gravity is a hundred times stronger, and the escape speed also happens to be a hundred times larger, a few hundred kilometres per second. At the surface of a white dwarf, gravity is a million times stronger than on earth, the escape speed is a thousand times greater, and the average density of a white dwarf is a million times that of water. What will matter be like at these phenomenal densities? Normally most of an atom consists of empty space, a tiny nucleus with a few slightly larger electrons in orbits of radius much larger than their size. The atoms in a white dwarf are so tightly packed together that the electrons no longer have space to orbit round their nuclei or even to move freely around as in an ionized gas. In this tightly packed state the electrons exert a strong pressure, known as the 'electron degeneracy pressure', and it is this pressure that holds the white dwarf up against gravity, pushing up on the outer layers of the star and stopping them from falling in.

Neutron stars

We travel now to a second bright X-ray source, Her X-1 in the constellation of Hercules. Its X-ray appearance differs from Sco X-1 in two ways. First we see that the X-ray source is eclipsed every 40 hours by a large object, the primary of the binary system, not detectable in X-rays. And the secondary on which the gas is pounding is much smaller, only 100 kilometres in size. We can deduce this by watching how long the primary takes to cover up the X-ray source. The gas hits the surface of the tiny secondary at a speed of thirty thousand kilometres per second, and the gravity there is ten thousand million times that on earth. We notice that the gas is striking a solid surface but this is not a rock-covered planet. The mass is similar to that of the sun and the average density is a million million times that of water. In the visible band we see a luminous blue star pouring out gas through its Roche point, but of the secondary there is no sign. Even when we look at the spectrum of the binary system we see only the characteristic lines of the primary star. Because of the Doppler shift these lines are changing wavelength to and fro as the blue star orbits around its unseen companion.

From its mass and size we deduce that the secondary X-ray emitting object is a neutron star, like the one we encountered at the centre of the pulsar in the Crab nebula. A neutron star is an even more condensed type of dead star than a white dwarf and is formed during a supernova explosion. The huge density of a neutron star means that the matter is almost entirely in the form of neutrons, which are crushed together so tightly that they cannot move around freely. The star is supported against gravity by the 'neutron degeneracy pressure', which is analogous to electron degeneracy pressure in white dwarfs except that now the electrons have all been combined with protons to give neutrons and it is the neutrons which are squashed together till they touch.

This cold compressed matter in a neutron star has some weird properties. The surface becomes a crystalline solid that conducts heat and electricity perfectly (a 'superconductor'). Below the solid crust the neutron fluid has no viscosity at all and if it is disturbed, the motions in it never die out as they do in a normal fluid (it is a 'superfluid').

The existence of these strange objects was deduced theoretically in the 1930s long before any observations could be used to support the idea. Had we been passing this binary system at the time of death of the star that became the neutron star, we would have witnessed one of the most spectacular events of the stellar landscape, a supernova. Whichever band we looked in we would see an enormous surge in brightness as the massive star blew up. Most of the matter of the star would have been blown off, but the inner core imploded to form the very dense neutron star. Some of the matter thrown off would have fallen on the primary star, and might even have increased its mass sufficiently to speed up its evolution. Originally the compact object (which is now the secondary) was probably the primary, more massive star of the binary system. Around the sky there should be many binary systems in which both stars have completed their evolution and are white dwarfs or neutron stars, but these would be very hard to find.

Sco X-1 and Her X-1 are stars in the stellar graveyard which have been lit up by gas overflowing from a companion star through its Roche point. There is another way that a dead star in a binary system can come to life in X-rays. As the primary, more massive, star nears the end of its own life, its stellar 'wind' may become something of a hurricane. Enough of this gas may fall on the secondary white dwarf or neutron star to light it up in X-rays. This is especially plausible for a class of X-ray sources where the primary is a very luminous blue giant star of ten to twenty times the mass of the sun. In one of these, Cen X-3 in Centaurus, the X-ray emission sometimes fades away, as if the stellar wind has faltered in its strength.

Black holes

For our third stop in the stellar graveyard, we travel towards the constellation of Cygnus, to Cyg X-1. And here we find the most famous of the discoveries of X-ray astronomy. To see a white dwarf in a binary system lit up in X-rays is interesting but could perhaps have been foreseen from what we knew about how stars evolve. To find a neutron star illuminated in the same way is more important. Although astronomers were sure when they found pulsars that a neutron star was responsible for the radio pulses they were seeing, they did not feel they were seeing the neutron star itself. In X-rays we feel we can really see gas raining down on the crystalline crust of these amazing stars.

At first sight the X-ray source Cyg X-1 resembles those we have already explored. We can see that the X-rays come from very hot gas in a very compact region. From a certain angle (not the one we see from the earth, though) the X-ray emission is eclipsed regularly by the primary star of the system. In the visible band we see this star is a highly evolved blue star which is on one of the periodic contractions from its normal red giant state in order to ignite some new nuclear fuel. The star is highly distorted in shape and is losing gas through its Roche point. When we spread the visible light from the system out into its constituent wavelengths and we find only one set of spectral lines, those of the primary, and we see them oscillating to and fro as the primary moves round its orbit. What dark companion is it orbiting round?

We can deduce quite a lot about this invisible object. From the period and radius of the orbit of the primary we can deduce the mass of the secondary star to be five to ten times as heavy as the sun. When we work out how massive a white dwarf or neutron star could be by considering how heavy a star could be held up against gravity by electron or neutron degeneracy pressure, we find that a white dwarf or neutron star could not be more than twice or three times as heavy as the sun. What happens when the remnant of a star after all evolution, mass loss and explosions, is heavier than this two or three solar mass limit? There are no nuclear fuels left to keep the centre of the star hot and provide a strong pressure to push outwards and oppose the force of gravity. Electron and neutron degeneracy pressures, which managed to hold white dwarfs and neutron stars up are inadequate in this case. There is no alternative for the remnant but to let gravity take its course. Now gravity is an insatiable force. Once it has an unresisting body in its grip, it never lets go. As the body gets smaller, gravity gets stronger still. Under gravity alone a body will collapse to a point.

We know now that the simple ideas about gravity that we have inherited from Newton fail when we want to consider very strong gravity. We have to turn instead to the General Theory of Relativity, invented by Albert Einstein in 1916 and now widely accepted by scientists as the right theory about gravity to use when Newton's theory starts to break down. Einstein's theory predicts many strange effects, like the bending of light, which have now been observed. However when it comes to the fate of collapsing object in the grip of gravity, the General Theory of Relativity tells us the same thing as Newton's theory. Gravity is inexorable. Once the collapse of matter is not resisted by sufficient pressure, the matter will collapse together to a state of virtually infinite density. General Relativity, however, does tell us one new thing, that this final 'singular' state will be hidden from view. At a certain point the speed with which the matter is falling inwards under gravity will reach the speed of light. From this point on we lose sight of the matter, for no light signals can reach us any more. The matter is hidden from us by a 'horizon'. We call it a 'black hole'. Light cannot escape from the horizon, although both light and matter can fall in. The heroic astronaut sent by NASA to investigate the black hole still receives his messages from Houston as he plunges towards destruction in the infinite gravity of the singularity, but of his despatches back not a word is heard.

Once we realize that the secondary of Cyg X-1 is so heavy that it can surely only be a black hole, we approach it with circumspection. The X-rays come from a whirling disc of gas surrounding the black hole, a whirlpool with blackness at its core. Gas striking the disc at very high speeds gets heated to the huge temperatures which produce the X-rays. At the inner edge of the disc the gas dribbles away into the black hole. Very rapid fluctuations in the X-ray brightness tell us that we are seeing turbulence close to the inner edge of the disc of gas.

The black hole in Cyg X-1 is one of the wonders of modern science, not expected perhaps even by many of those who believed in Einstein's General Theory of Relativity. Many massive stars may end their life in this way and the cosmic landscape may have many tiny dark spots on it. Yet these are only drawn to our attention if matter falls into them and by colliding with other matter radiates some of the energy of its motion. If the core of the earth collapsed down to a black hole we would hardly notice the difference. The matter inside a black hole still exerts its normal gravitational influence, so life would go on as usual provided the crust held together. In principle there could be very tiny black holes which pass right through the earth like a very energetic cosmic ray. It has even been suggested that the Siberian meteorite explosion of 1908 was due to a small black hole, but this is not very likely. It has to be admitted that it will always be hard to prove that any particular phenomenon, including Cyg X-1, is really due to a black hole.

Other X-ray sources in the Milky Way

As we travel round the Milky Way we find examples of all three types of X-ray source, where the secondary star of the binary system is a white dwarf, neutron star, or black hole. We also see 'soft', lower frequency X-ray emission from the thin gas at a million degrees spread between the stars which we found on our ultraviolet voyage and this is enhanced when we look in the direction of supernova remnants. We see too soft X-rays from the coronae of near-by stars. The bright star Capella, in the constellation of Auriga, the Charioteer, has a corona ten times hotter than the sun, for example. The sun's mild spottiness pales into insignificance compared with what we see in the binary system RS Canus Venaticorum (the Hunting Dogs). This consists of two stars similar to the sun in a close orbit round each other every few days. The cooler of the two stars has 40 per cent of one hemisphere completely covered with 'starspots' (named by analogy with sunspots). Stars like this are similar to the active sun but hundreds of times more active and they are strong sources of soft X-rays.

An example of accretion of gas by a white dwarf in a binary system, at a level much less dramatic than Sco X-1, is the 'dwarf nova' U Geminorum, in the constellation of the Twins. Every so often this flares up in the visible band as gas from its companion M dwarf overflows its Roche lobe and falls onto the magnetized white dwarf. This gas gets heated to very high temperatures, so U Geminorum can be seen as an X-ray source during an outburst.

We travel beyond the limits of our Galaxy. In other near-by galaxies we see similar types of X-ray source. In every direction we see too a general bath of X-radiation, of the same intensity whichever direction we look. The origin of this X-ray 'back-ground' radiation remains a mystery. Perhaps it is made up of many sources too faint to see individually.

Clusters of galaxies

Our X-ray eye is drawn towards the huge cloud of galaxies in the constellation of Virgo, the Virgo cluster, fifty million light-years from earth. The inner part of the cluster appears as a huge cloud of X-rays, centred on the massive galaxy M87. The X-ray light does not come from individual galaxies, but from hot gas spread through the cluster. This gas makes up about a tenth of the mass of the whole cluster and seems to be heated by collisions between clouds of the gas as they orbit around at random in the cluster's gravity. We often find these cluster X-ray sources when a cluster is rich enough in galaxies. Some of the gas probably comes from spiral galaxies that have collided with each other. In such an encounter the star systems pass right through each other without having much effect because the stars in a galaxy are so far apart compared with their size, but the clouds which make up the gaseous discs of spirals collide violently and get left behind the two moving galaxies. When this happens to a spiral galaxy, star formation ceases because there is no gas left to form new stars, and the galaxy gradually becomes redder with time as the more massive, bluer stars evolve and die. The spiral arms disappear and we can no longer see much difference between the stars of the halo and the stars of the disc. This gas-denuded spiral there-fore ends up looking pretty much like a red elliptical, with no young blue stars. However we can still notice that it is lens-shaped rather than genuinely elliptical if we look carefully at the distribution of visible light across the galaxy. These encounters between galaxies in rich clusters explain why such clusters tend to be rich in ellipticals (and lens-shaped galaxies), whereas in the general field spirals predominate. When we travel out to the most distant clusters discovered so far, we do indeed find a surpris-ingly large proportion of blue galaxies, presumably spirals, compared with any near-by cluster like Coma or Virgo. Once there is a reasonable density of gas in a cluster, spirals may also have their gas stripped away as they plough through the gas.

Some of the X-ray-emitting gas could be left over from the time when the galaxies in the cluster formed. We don't yet know whether a cluster of galaxies forms as a huge cloud of gas out of which galaxies condense, or whether the galaxies form first, independently, but then congregate together into a cluster. If the cluster formed first as a cloud of gas, we would not expect the condensation of galaxies to be a hundred per cent efficient and some gas should be left over to be pushed this way and that by the galaxies. If the galaxies formed first and then congregated into clusters, we would not expect much gas between them to start with and the X-ray-emitting gas must have come from galaxy collisions.

An important clue to the origin of this gas is found if we look at these cluster X-ray sources in X-ray light of a very particular wavelength, 1.8 Å. This is one of the characteristic wavelengths of ionized iron with 24 of its 26 electrons stripped off. The abundance of iron in the cluster gas turns out to be almost the same as that in the gas and stars of the Milky Way and other galaxies. Now iron is made in stars, so this iron emission tells us that the cluster gas cannot be virgin, primordial gas. Much of the cluster gas has been inside stars at some time in its history and therefore presumably inside galaxies. So the theories where gas is stripped out of galaxies must be at least part of the truth.

This gas in clusters makes itself felt in the radio band too. When we see a galaxy with a powerful double radio-source in a cluster of galaxies, it often looks rather lopsided. Instead of finding the usual pair of radio sources symmetrically on either side of the galaxy, they have been swept sideways into a V-shaped configuration, with the galaxy at the vertex of the V. The radio-emitting clouds of fast-moving particles seem to have been swept backwards by the galaxy's motion through the gas in the cluster, which we presume to be in the direction the V is pointing. These have been called 'head–tail' radio sources, since there is often radio emission from the galaxy itself, which then looks like the head of a double-tailed tadpole. They also look a bit like the wake of a ship ploughing through the ocean.

M87 and other active galaxy nuclei

While we are in the Virgo cluster we notice that the nucleus of the prominent radio galaxy M87, which we have encountered several times already on our travels, contains a compact X-ray source right at its core. This X-ray source does not have the signature of thermal radiation from hot gas — that it is emitted mainly at frequencies such that the energy of the photons is roughly equal to the energy in the individual atoms in the gas. Instead we see emission from this core all through the X- and gamma-ray band, which must therefore originate in a 'non-thermal' process. One possibility is synchrotron radiation which we met on our earlier voyages — radiation from electrons moving through a magnetic field at speeds close to that of light itself. Another type of 'non-thermal' process which may be important in the X-ray band is called 'Compton scattering', after the American physicist A. H. Compton who discovered it in 1923. An electron moving close to the speed of light collides with a photon of low energy, say from the radio or infrared band, and gives some of its energy to the photon, boosting the photon's energy and turning it into an X-ray photon.

We realize that our band is exceptionally suitable for investigating the activity in the centres of galaxies which we discovered in the radio and ultraviolet bands. With X-rays we are probing down to regions in the galaxy nucleus close to the energy source itself, not much larger than our solar system. We travel to the nucleus of the Seyfert galaxy, NGC 4151, that miniature quasar in the heart of an otherwise normal spiral galaxy which we encountered on our last voyage. Here too we find strong non-thermal X-ray emission from the compact core. At the hard X- and gamma-ray end of our band we also see some excess radiation which may be due to gas at a hundred million degrees. As we travel to other near-by Seyfert galaxies, we find that they too are strong X-ray emitters. This X-ray-emitting characteristic was shown conclusively only recently by the *Arial V* satellite in 1978, although X-ray emission from some Seyferts was seen by the pioneering *Uhuru* satellite.

We travel on towards a more dramatically active galaxy, the quasar 3C273. It is no surprise to find that this active galaxy is emitting X-rays strongly. Other quasi-stellar radio sources are X-ray emitters as well but they are harder to see since they are usually so much further away.

What is the powerhouse for an active galactic nucleus? It has to put out a thousand times as much energy as the total starlight of the galaxy for bursts of up to a million years to fuel the quasar phase and it has to work at a weaker level for much longer periods to supply the energy for radio galaxies and Seyfert galaxies. Recall that when we travelled to the nucleus of M87 in the visible band we saw that the star motions there pointed to a dark compact mass of a thousand million suns. Could this be a monster black hole? Many astronomers who work on quasars and active galaxies think that it is indeed massive black holes that provide the powerhouse for these objects, but the nucleus of M87 is the nearest we have to direct evidence. Other possible models can explain transient outbursts, for example collisions between stars in a very dense star cluster or gigantic flares in a massive, compact, magnetized gas cloud. But the great advantage of a black hole is that it can flare up at any time that it is fed by gas from outside and can produce activity at a wide range of powers. Of course we do not feed the monster with gas straight down its throat. For the falling gas to give out some energy it must collide with something else before it passes the dreaded horizon of no return. We need something like the structure round the black-hole X-ray source, Cyg X-1, but on a much larger scale. The black hole needs to be spinning and the gas that falls in needs to build up a rotating disc of gas outside the black-hole horizon. Then when new gas falls in, it can collide with gas in the disc and get heated up to X-ray temperatures of hundreds of millions of degrees. Some of the energy of the in-falling gas can go into accelerating particles to near the speed of light to provide the radio and other radiations.

Gamma rays and the Vela bursts

We have hardly mentioned the high frequency, gamma-ray end of our band. Very few gamma-ray sources have been found to date, about two dozen in all, most by the European *COS-B* satellite, launched in 1975. The resolving power of the telescope in this satellite is very poor and it is hard to be certain what it is pointing at, but the Crab pulsar and the quasar 3C273 are two of the familiar objects whose non-thermal radiation has been traced out to the very high frequencies of the gamma-ray band. We see gamma-ray emission from the Milky Way, which is produced by the decay of elementary particles called 'pions' created by high energy cosmic rays ploughing through the interstellar gas. Apart from 3C273 and a few pulsars, most of the gamma-ray sources remain unidentified.

Astronomers were startled a few years ago when the US H-bomb laboratory at Los Alamos announced that they had a network of gamma-ray satellites in orbit, which they used to monitor any clandestine nuclear explosions in space as part of the SALT agreements. Intense gamma-ray radiation is one of the main forms of energy which emerges from a nuclear blast. With this *Vela* network of satellites, as it is called, sudden bursts of gamma radiation had been detected. From the fact that the bursts appear in the same direction on the sky to two widely spaced satellites, it can be deduced that the *Vela* bursts come from far outside the solar system. Seven or eight bursts are seen each year and their origin is not known at all. Perhaps they are due to nuclear reactions taking place in gas which has fallen onto the surface of a neutron star; a kind of gamma-ray nova.

Gamma-ray spectral-line emission has been seen in solar flare events and perhaps from the centre of our Galaxy. These lines arise from nuclear reactions; for example, the annihilation of a positron by an electron or the capture of a neutron by a hydrogen atom to make deuterium, or from transitions of an atomic nucleus from one excited state to another.

Gamma-ray astronomy is still in its infancy, despite many years of pioneering work with balloons and satellites. However it has great potential for astronomy, both for studying nuclear processes and in understanding active galaxies, some of which may have their main energy output in this band.

6
FIFTH VOYAGE
The infrared landscape

The infrared band and its windows on the cosmic landscape

We exchange the hostile landscape of the X- and gamma-ray frequencies for the warmth of the infrared. How welcome after the cold night are these warm invisible rays of the sun. This is the band in which the earth radiates away most of the energy it has received from the sun, all that has not been used by the growth of plants, in the driving of the weather, the winds, the oceans. Radiating in this band too is the human body, at its normal temperature of 37 °C. Infrared light is familiar to us because although our eyes are blind to it we can feel it warming up our bodies. Our bodies, however, make rather limited detectors of the infrared cosmic landscape, able to detect only the sun. Rattlesnakes have a better impression of the infrared landscape as they are equipped with an extra pair of 'eyes' especially for this purpose. They see the warm bodies of their prey and its warm footprints standing out against a bright landscape — brightest where the landscape is warm and darkest where it is cool.

Astronomers have had to develop their own infrared 'eyes'. In the near infrared the lead-sulphide cell, similar to the cadmium sulphide cell used in camera light-meters, has been very effective. For longer wavelengths astronomers use detectors which consist of a germanium crystal whose electrical resistance is very sensitive to heat. This kind of heat detector or 'bolometer' is sensitive to the whole range of infrared wavelengths, so to examine the infrared sky in light of a particular wavelength we have to put a filter in front of the detector which allows through only a small range of wavelengths. Another kind of detector used in the near infrared is the 'photoconductor' — infrared light falling on a thin metallic film causes a current to flow which can be measured.

The existence of invisible radiation from the sun was first demonstrated by William Herschel in 1800. He formed a spectrum of the sun's radiation with a prism and then held a thermometer in the blank region beyond the red end of the spectrum. The mercury started to shoot up the thermometer, showing that the sun emits radiation with wavelength greater than that of red light. Thus began the astronomy of the invisible wavelengths. The next cosmic source to be detected in the infrared was the moon, detected in 1856 by the Scottish Astronomer Royal, Piazzi Smyth, during an expedition to the island of Tenerife. Measurements of some planets and stars were made in the 1920s, but the field did not start to expand rapidly until the 1950s and 1960s. Our infrared band stretches from wavelengths of just under one micrometre (a thousandth of a millimetre) to 300 micrometres. From the ground though we only get brief glimpses of the infrared cosmic landscape. Most of the light in our band is heavily absorbed by the molecules of the earth's atmosphere, especially carbon dioxide and water.

Although the earth's atmosphere is like an impenetrable wall for most infrared wavelengths, the landscape peeps through a few wavelength 'windows'. These are small ranges of wavelengths which the molecules of air happen not to be so effective at absorbing. The atmosphere itself radiates in the infrared and if we were to spread the wavelengths out with a prism into a spectrum, we would see a forest of lines across almost all parts of the spectrum due to the characteristic wavelengths of the molecules of air. The windows are chance gaps in the forest, for the atmosphere's absorption takes place at these characteristic wavelengths.

At sea level we can see through several windows in the near infrared, between 1 and 3 micrometres, but not much else. The window at 2 micrometres was used by two California Institute of Technology astronomers, Gerry Neugebauer and Bob Leighton, who surveyed the whole northern sky during the 1960s. They found nearly 6000 sources of near-infrared radiation, almost all of them stars but many of them previously undetected in the visible band because of the shroud of dust in which they are immersed.

From a mountain in a dry climate other windows at 5, 10, and 20 micrometres become usable, but to see the landscape of 20 to 300 micrometres or to see between the ground-based wavelength windows, we have to take to high-altitude balloons or aircraft. The Kuiper Airborne Observatory, a 1-metre diameter telescope on a C 141 transport aircraft, has made a tremendous impact on these wavelengths. Towards dusk it takes off from the NASA–

Ames airfield at Mountain View, California, and travels across the US on a path determined by the sources the telescope is to look at. Once off the ground the plane is put on auto-pilot, connected via the onboard computer to the telescope. The plane is therefore flown by the telescope!

Rockets have been used to survey the sky at 4, 11, and 20 micrometres, and a joint US–Dutch–UK infrared survey satellite, *IRAS*, is being built as I write. The military already have infrared satellites to monitor enemy spacecraft and they have probably surveyed much of the cosmic landscape by accident. However, the results of these surveys have not yet been released.

Since our band has not yet been adequately surveyed at the longer wavelengths, our picture of the landscape is very incomplete. The infrared may have yet to find its quasars, pulsars, or black holes, its own special great discovery. Yet when the *IRAS* satellite has flown and made its survey, an era will have come to an end. The whole cosmic landscape from radio to X-rays will have been searched in some detail. A century may pass before we understand all the phenomena which have been revealed to us by the opening up of the invisible wavelengths for astronomy. But there cannot again be the experience of looking at the sky for the first time in a new wave band. For me it has been a wonderful time to be alive and to be an astronomer. And so I try to share this experience with you.

The solar system

We set sail on our fifth voyage. Although the sun has its peak brightness at yellow wavelengths there is still enough power at infrared wavelengths to make it the brightest object in the infrared sky. But if we had infrared eyes like the rattlesnake, would the rest of the sky seem light or dark at infrared wavelengths? We recall that the daytime sky is bright in the visible band because of scattering of blue and violet light by dust and molecules of air. Infrared light, on the other hand, is hardly scattered at all in the atmosphere, so the sky would be dark day and night in our band but for one thing. The air itself radiates strongly in the infrared band and all but drowns out the faint signals from the cosmic landscape reaching earth's surface. To see the tiny extra contribution from a cosmic source, the radiation from a near-by direction has to be subtracted from the radiation in the direction of the source. The atmospheric contribution should then cancel leaving the cosmic signal.

The planets are exceptionally prominent in our band and we see them by virtue of their re-radiation of the light they have absorbed from the sun, rather than by reflected light as in the visible band. They have average temperatures ranging from 200 °C for Mercury and 300 °C for Venus to only –220 °C for Pluto. Gaseous or solid bodies in this temperature range radiate most of their energy in the infrared. The infrared is above all a band of warm and cool bodies, one at 500 °C radiating most strongly in the near infrared while one at –250 °C, only 20 degrees above absolute zero, emits most strongly in the far infrared at a wavelength of about 300 micrometres. Absolute zero is the temperature at which the motions of the atoms in matter cease — no matter can be colder than this. The human body, at 37 °C, radiates most strongly at a wavelength of around 10 micrometres, and this is also where the earth has its strongest emission.

The high temperature of Venus was quite a puzzle at first, since it seemed to be hotter than we would expect for a planetary object at that distance from the sun. This is true on a less dramatic scale for the earth too and the explanation lies in the so-called 'green-house' effect. Because carbon dioxide and other molecules are such effective absorbers of infrared radiation, the earth's atmosphere is better at letting the radiation from the sun in (this is mostly visible light) than it is at letting the earth's infrared radiation out. The earth therefore gets warmed up a bit. This happens even more dramatically on Venus, which Russian and American spacecraft have shown to be at a temperature of 480 °C on its rocky and cratered surface. It is also from the infrared, through a strong absorption at a characteristic wavelength of 11 micrometres, that we learn of another strange feature of the Venusian landscape, the presence of sulphuric-acid droplets in the atmosphere. The prospect of sitting on the boiling hot rocks of Venus, being rained on with sulphuric acid, is not an attractive one.

By studying the infrared radiation from the moon, it was possible to deduce in 1948, long in advance of the moon landings, that the moon was covered not with rock or lava but with a fine powder. At different infrared wavelengths our eyes penetrate to different depths below the surface and so we see slightly different temperatures. The variation of infrared colour with wavelength is therefore a good clue to the nature of the surface of a solid planet. The technique can also be used to tell how well crops are growing on the earth and this has peaceful applications, as well as being part of the work of the spy satellites.

The infrared is the wavelength band where the matter of the solar system gives back the energy it has absorbed from sunlight. The disc-shaped cloud of interplanetary dust which is responsible for the zodiacal light glows in the infrared. The dust appears to be made of iron and magnesium silicates like the rocks on earth and the particles range in size from a tenth of a micrometre through to sand-sized particles and up to rocks, boulders, and asteroids. The larger objects like asteroids are not nearly as common as the smaller rocks and grains, but several thousand of them are known from searches in visible light. As sensitivities in the infrared band improve, hundreds of thousands of asteroids may be detected, and we might eventually hope to see some of the billions of comets from the cometary cloud beyond Pluto. We can already see in the infrared those rare but dramatic comets that plunge in towards the sun as a result of some chance gravitational deflection by a star or planet.

Red giants and dust shells

We travel onwards into the landscape of stars. In the near infrared part of our band we see a sky of red stars, dying red giants, and low-mass red dwarfs quietly burning hydrogen for aeons to come. Already some new objects have crept into view. On investigation we find a red giant star hidden inside a dense cloud of dust. A large proportion of red giants that are slowly variable like Mira turn out to be surrounded by a cloud of dust which radiates in the infrared. The star pulsates irregularly and throws off part of its surface layers. When this has expanded a little way from the star and cooled down, grains of dust condense out. These absorb some of the starlight and re-radiate the energy in the infrared. Where the dust cloud becomes very thick the star becomes too faint to be noticed in the visible band and only the infrared radiation escapes.

Most of the stars that we can see in our band surrounded by dust are in the red-giant phase, but they fall into two subgroups. Some have oxygen-rich surface layers and the grains that condense out of their cooling, expanding gas shells are composed of silicates. The second group have carbon-rich layers and the grains that form round them are probably graphite or silicon carbide. Silicates are especially efficient absorbers and emitters of light of around 10 micrometres in wavelength and so infrared light from a silicate dust shell is usually either especially strong or weak at 10 micrometres, depending on whether emission by the dust or absorption of starlight by dust is more important. These dust shells around red giant stars seem to be the main place where the grains of interstellar dust are formed.

The dust clouds where stars are forming

As we look out into the infrared landscape we notice how much clearer and more transparent the Milky Way seems. We can see right through the dark clouds that obscured our view in the visible band. The concentration of stars in the Galactic nucleus stands our clearly around the constellation of Sagittarius. We notice that the same dark clouds are themselves radiating in the far infrared part of our band confirming that they contain dust. We can also see infrared radiation from dust spread out between the stars and dark clouds.

However these dark clouds are by no means the most prominent nebulous objects in the far infrared. We travel to Orion the hunter, to the nebula in the middle of his sword. Not far from the four stars that make up a group called the Trapezium, in an area of the nebula not especially remarkable in the visible band, we see a very intense ridge of far-infrared radiation. In the middle of this there are two sources which are prominent in the middle and near infrared. One, the Kleinmann–Low nebula, is slightly extended and may be a cloud of dust surrounding one or more completely obscured stars. The other, the 'Becklin–Neugebauer object' (named, like the nebula, after its two co-discoverers) is far more compact and is probably a star being born. If the latter source is interpreted as a single warm body, it would have a temperature of 200 °C, much cooler than the coolest stars known. However it might also be interpreted as a much hotter object, heavily reddened and obscured by a shell of dust. The far-infrared ridge is due to radiation from dust heated by the central sources to temperatures within the range –240 to +200 °C.

What are the hidden sources of heat in the Orion dust cloud? Most probably they are young massive stars which in due course will look very much like the Trapezium stars, blue giants. At the moment the stars may still be forming and may not yet have reached a steady hydrogen-burning state. Or if they are already blue hydrogen-burning stars they are not only shrouded in dust but are also surrounded by gas so dense that the stars have not made much progress in forming the cloud of ionized gas which usually surrounds such stars. The Orion nebula, which we saw in the visible and radio band, is just such an ionized cloud created by the Trapezium stars. The near-by infrared radiation marks the site of a cloud of gas and dust still in the process of forming into stars. The associated gas will become visible when we make our voyage in the microwave band. From the far-infrared emission we can deduce that the density of this cloud is a thousand times greater than the typical interstellar cloud of neutral hydrogen. As we travel round the Milky Way we find many examples of these dense clouds of gas and dust radiating in the far infrared, illuminated by invisible sources embedded deep in the dust. In almost all cases that we see strong far-infrared emission we also see young blue stars and their associated clouds of ionized gas.

The far-infrared landscape is therefore dominated by regions where new stars are forming out of clouds of dust and gas. We see these events secondhand, however, via radiation absorbed and re-radiated by grains of dust. At first the dust blots out the stars completely in the visible range. The stars only start to peep out there when their light has pushed the placental gas and dust away from their vicinity. We recall from our visible voyage the nebulosity surrounding the Pleiades where we could see this process in its last stages.

Naturally it is easier to see the emerging stars if they are massive and luminous. Most of the clouds we notice in the far infrared are those in which a massive blue star, or cluster of such stars, is heating up the dust. To find a youthful version of a more modest star like the sun, we travel to the nearest dark cloud of all, a large dust cloud about three hundred light-years away in the direction of the constellation of Taurus. Embedded in this we see some rather unusual red stars, named after the star T Tauri which was the first noticed. Normally when we spread the visible light of a red giant or dwarf out into a spectrum we see dark bands across it due to carbon monoxide and titanium oxide. These tell us that we are looking at a relatively cool star, with a temperature of two or three thousand degrees. However the spectra of the T

Tauri stars show in addition emission lines of hydrogen and helium characteristic of a much hotter gas. We find that, like the similar lines in the sun's spectrum, they are arising from a zone of hot gas above the visible surface of the star. We also notice that the lines are shifted in wavelength due to the motion of the gas. Sometimes gas is falling onto the star and sometimes the star is ejecting a strong stellar wind.

Although young, these stars are quite advanced in their development. Already they are contracting and getting hotter at the centre. They are not yet hot enough at their centres to burn hydrogen, but they have a stable structure and it is inevitable that they will become normal hydrogen-burning stars. In this phase, which lasts about a hundred million years, the sun would be much larger, much redder, and more luminous that it is now. It is odd that even before a star like the sun is completely formed, it already spends much of its time copiously blowing off gas from its surface.

To find stars at an earlier stage still we must look for cooler objects radiating entirely in the infrared. We do find such objects within the dense star-forming clouds we have already explored, for example the Becklin–Neugebauer object, but we cannot be sure whether they are stars already formed but hidden in dust or the protostars we are searching for.

The life cycle of stars and of atoms

Even though some parts of the story are still mysterious to us, we are beginning to have a picture of the life cycle of stars and gas. A cloud of gas starts to collapse and condenses into stars. To start with the stars are shrouded in dust but they gradually burn their way out, melting the dust nearest to them and driving the rest away. We have a cluster of stars. The more massive stars burn their hydrogen rapidly, and start to burn other fuels, becoming red giants in the process. All the while they have been blowing off a wind of gas, and in their last stages they blow off a substantial part of their mass as supernovae, leaving behind a neutron star or black-hole remnant. The gas that has blown off the star is blown around by winds from other stars and collects into small clouds. These little clouds coalesce until a large cloud is formed. The cycle is ready to begin again.

We saw in the visible band that the young recently formed stars lie along spiral arms in galaxies like our own and the Andromeda nebula. These represent the current wave of star formation in the disc of the galaxy. This wave is driven by an instability in the gravitational field of the stars weaving in and out of the disc. The spiral pattern rotates through the disc and the gas clouds experience a sudden compression as they pass through it. This is enough to trigger the most massive of them to collapse and condense into stars.

What of the dust grains that are so important for our infrared band? In the infrared light of most dust cloud sources we see an excess or deficit of light of wavelength around 10 micrometres and deduce that silicates are present. From another feature at wavelengths of around 3 micrometres we can deduce that the grains are often coated with ice where conditions are cold enough. They probably have a more complex composition though, containing most of the elements of the periodic table. The abundance of the elements is remarkably similar in different parts of our Galaxy, apart from the ratio of hydrogen and helium to the rest. The surface layers of the sun, meteorites, distant hot clouds of gas, all seem to have this 'cosmic' relative abundance of heavy elements. This must reflect the way these elements are formed in nuclear reactions in stars and then blown out in winds and supernovae. In the cooler parts of the Galaxy most of the heavy elements (everything except hydrogen and helium) is locked up in these tiny grains. We will return to this question on our final voyage when we investigate the world of interstellar molecules.

Let us follow the fate of one particular heavy atom, of carbon say, from the moment that it is first thrown into interstellar space in some dramatic stellar fatality. It spends hundreds of millions of years locked up in a grain a tenth of a micrometre in size, absorbing and scattering any visible or ultraviolet light that passes. The grain passes most of this time at a very cold temperature, only 10 degrees above absolute zero (i.e. at –263 °C), radiating feebly in the far infrared. The grain is blown around by the prevailing starlight and its motion is resisted by any local gas. Sometimes the grain finds itself in a small cloud of gas, but it may be blown out by the pressure of starlight. Eventually it finds itself in the interior of a moderately thick cloud, where starlight cannot penetrate. The cloud sweeps up more material and increases in density. The shock of the spiral density wave rotating round the galaxy and compressing all the gas it meets, passes many times, and eventually jerks the cloud into collapse. Our grain finds itself inside one particular concentration of gas which is to form a star. As the fragment condenses and heats up, the grain is melted and our original atom finds itself moving around freely in a hot gas again. As the star starts to fuse hydrogen, our atom may find itself in the nuclear burning zone acting as a catalyst. If the star is of low mass our atom may remain locked up here for hundreds of thousands of millions of years. In a massive star the atom may find itself being blown out of the star again after only a few million years. Once in its many cycles of life, it may find itself not in a star but in a planet. It might even find itself within my hand or your eye.

The origin of the solar system

Although most of the dust mass of our Galaxy is in the form of very small grains, some of it takes the form of much larger grains of dust. We are sitting on one now, the earth. The solar system contains large icy grains like Pluto, large silicate–iron grains like Mercury and the moon, and large silicate–iron grains partially coated with ices, like the earth and Mars. In the form of inter-planetary dust, the solar system also contains much smaller grains of course.

Why does the solar system have a significant proportion of its mass in very large grains? The crucial distinction between the solar system and an interstellar cloud lies in the very high density phase that the solar system material has been through during the formation of the sun. The detailed scenarios for the formation of the planets are so diverse that it is hard to feel much progress in understanding has been made at all. One theory involves chemical separation in a disc-shaped cloud of gas from which the sun forms in the centre and the planets in different zones further out. A second theory involves the gradual building up of the planets from much smaller floccules of matter through collisions and capture. And a third theory depends on the capture of material by the sun from a passing protostellar cloud. In each case much of the material not incorporated into planets is swept away during the sun's 'T Tauri' phase. It is hard to choose between these theories when we know of only one example of a planetary system, though most workers in the field favour theories in which the planets formed as part of the process by which the sun itself formed. However there is still no agreement whether planets form around most stars or only in some specialized cases, for example only where a single star rather than a binary or multiple star system forms.

As we travel through the infrared cosmic landscape, can we see any signs of planets forming elsewhere? The nearest direct evidence we can find is certain very young stars surrounded by a disc of dust which radiates in the infrared. Perhaps this disc could eventually coalesce into planets. From the almost imperceptible wanderings of several near-by stars we can deduce that they have small companions, but the masses of the companions deduced in this way are, with one exception, one or two per cent of our sun's mass, i.e. 10–20 times the mass of Jupiter. Such objects could in fact be tiny stars, rather than planets, for they may be undergoing nuclear reactions in their core. The one exception is Barnard's star, the next nearest to the sun after the α-Centauri system, 5 light-years away. It has been claimed that this star has one or two companions of mass about that of Jupiter. This is still a matter of dispute between astronomers. It is an act of faith, based on rather shaky probabilistic arguments, to say that other planets like earth exist in the universe. This is a question I will return to in the last chapter.

The nucleus of our Galaxy and other galaxies

We travel onwards, out of the plane of the Milky Way. Below us we see the general infrared glow of the interstellar dust, the blotchy structure of the dark clouds and then the bright dense clouds illuminated by young stars and protostars tracing out the spiral arms. Towards the centre of our Galaxy we see an exceptionally bright cloud a thousand light-years across, and we can see within it the red giants which are illuminating it. Any quasar-like source in the nucleus of our Galaxy must be exceptionally weak. However there is just a hint of such activity from observations made in our wave band. Ionized neon has a characteristic wavelength of 12.8 micrometres. At this wavelength we can examine ionized clouds of gas very close to the Galactic centre. As in M87 we seem to see exceptionally high velocity clouds near the centre, pointing to a hidden, dark object of mass about ten million times that of the sun. Could this be a black hole again? If we switch to the radio band we can see a very tiny radio source here, of diameter only ten times that of the earth's orbit round the sun. However this source is giving off only about as much energy as the sun, a trivial amount compared with some of the other active nuclei we have seen in radio galaxies and quasars. The absence of any more violent activity in our Galactic nucleus implies that it is a hungry black hole, with little gas being fed into it at the moment.

As we look out towards other near-by spiral galaxies we see strong infrared emission from their nuclei and some of the brighter individual clouds in their discs. However certain galaxies stand out in the infrared cosmic landscape. In the nearest group of galaxies to us outside our own Local Group of galaxies lies the brightest far-infrared source on the sky outside the plane of the Milky Way, in the centre of M82. The nucleus of this galaxy is radiating almost as much energy in the far infrared as the Milky Way is radiating in starlight altogether. As usual this radiation comes from heated dust and the source of this heating is hidden from view. It could either be large numbers of massive stars from a recent burst of star formation in the nucleus or a non-thermal core, a miniature quasar, powered by a giant black hole.

M82 has a very unusual appearance in the visible band, with an irregular, cigar-shaped outline. Extending out from either side of the nucleus is a cone of bright streamers and filaments, very suggestive of a huge explosion in the nucleus. However since the central regions of the galaxy are so heavily shrouded in dust it is hard to be sure this is not an artefact of scattered light from dust filaments. One theory is that M82 has run into a huge inter-galactic cloud of gas and dust.

Seyferts, quasars, and dust

We travel on through the landscape of galaxies. We notice that the Seyfert galaxies, which were so prominent in the ultraviolet and X-ray bands, are also much stronger in the infrared than normal galaxies. In the near infrared this is due to the compact core which produces the non-thermal visible and ultraviolet radiation. This non-thermal radiation extends over a wide range of wavelengths and so starts to dominate over starlight (which is peaked in the visible band) once we get out of the visible band at either end into the ultraviolet and infrared. At longer wavelengths still, in the far infrared, we see emission from dust, coming from a region thousands of light-years in size. What has happened is that some of the ultraviolet and visible light has been absorbed by the dust and is being re-emitted in our band.

One of the nearest Seyfert galaxies, NGC1068, is perhaps the most remarkable galaxy of all in the far infrared. The amount of energy it is emitting in the far infrared exceeds by a factor of ten the total output of the galaxy in the form of starlight. If this energy were emerging in the form of a compact non-thermal visible and ultraviolet source then this galaxy would be the nearest quasar, since the compact nuclear source would blind us to the rest of the galaxy. Is this a quasar hidden in a cloud of dust? Will it eventually shine out as a true quasar?

We come to the bright quasar 3C273. It is a strong infrared source, but for once we do not see much sign of dust. All the way from the radio band to X-rays we see non-thermal emission from particles moving close to the speed of light.

We travel outwards and back in time. The galaxies become younger and have less heavy elements and hence less dust in them, so the infrared emission from dust round a newly formed star gets weaker. However in their youth galaxies formed stars far more rapidly than at present, so their total far-infrared output was probably larger than that we observe now.

We notice the growing intensity of the background radiation and that it moves to ever shorter wavelengths as we travel back in time. What is this background radiation? We must return to earth and change to the last of our wavelength bands to understand this.

7

SIXTH VOYAGE
The microwave landscape

The microwave band and its limitations

The cosmic landscape has been transformed before our eyes by
our voyages yet there are two fundamental questions that we
have not answered. How was this vast landscape of galaxies
formed? And how did life and we ourselves come to exist here on
earth? To find some answers to these questions we embark on our
sixth and last voyage in the microwave band.

We shall define our band as stretching from wavelengths of
about 300 micrometres (0.3 mm) to about 3 centimetres. At the
short wavelength, high frequency end of the band, it merges into
the far infrared, and at the long wavelength, low frequency end,
it merges into the radio band. The boundary between these bands
is not clearly defined and the observing techniques overlap too.
Radio receivers are used for wavelengths longer than about 2
millimetres while heat-measuring bolometers like those used in
the infrared take over for the shorter wavelengths. The name
microwave was coined in the 1930s for radio wavlengths shorter
than about 30 centimetres. Microwaves are familiar in two applications, the microwave oven and the microwave radio link, used
for intercity telecommunications. Microwave radio dishes and
transmitters can be seen on high buildings and hills throughout
the world, but they are usually operating at wavelengths slightly
longer than those of our band.

The microwave band is severely affected by absorption by
molecules of air. Wavelengths longer than 2 centimetres reach
the ground reasonably unaffected, but the shorter wavelengths
start to be severely absorbed. There are good windows at
1 centimetre and at 3 millimetres through which astronomy can
be done from sea-level sites. A good mountain site allows work in
windows at 1 and 2 millimetres to go ahead with some difficulty.
Very dry conditions can even allow some light through another
window at 800 micrometres. A dry, very high site, like White
Mountain, California, or Mauna Kea, Hawaii, both above 13 000

feet, allows the 400-micrometre window to be used. Here the astronomer has to cope with violent mountain storms that rage suddenly, and with the reduced oxygen of high altitudes. To observe outside the wavelength windows we need to take to a balloon-based platform or use a high-altitude aircraft.

The microwave band has suffered from a severe shortage of telescopes designed specifically for these wavelengths. For many years the 36-foot millimetre-wave dish operated by the US National Radio Astronomy Observatory on Kitt Peak mountain, Arizona, was the only large telescope dedicated to this band. Astronomers working at submillimetre wavelengths have had to use the existing optical telescopes on mountain sites, with severe competition for observing time from optical and infrared astronomers. The situation is improving now, with a profusion of large millimetre-wave telescopes being completed or built throughout the world.

One of the consequences of the shortage of telescope time is that no survey of the sky has been attempted in our band. For the first time on our voyages we are unable to look round the sky and see the cosmic landscape as a whole in our band. Instead selected points on the sky have been explored, mainly those from which strong radio or infrared emission comes. This is a technique which by its nature is not going to find new sources which are prominent in this band and no other.

We continue to see the non-thermal radio emission from radio galaxies and quasars extending into our band, with the main emission coming from very tiny sources in the galaxy nuclei. We see the radio free–free emission from hot clouds of ionized gas associated with young blue stars. We see, weakly, the extension of the far-infrared emission from dust in dense clouds illuminated by newly formed stars.

When we search over the whole microwave and submillimetre-wave landscape with sensitive detectors on satellites we shall be able to see the cold clouds of gas and dust spread between the stars of the Milky Way. We may even find cold clouds spread between the galaxies although intergalactic gas is more likely to be very hot and radiate in X-rays. Whether other unsuspected features will appear in the landscape remains to be seen.

The cosmic microwave background

However one unexpected feature of the microwave cosmic

landscape, found in 1965, has turned out to be one of the most significant discoveries of twentieth-century astronomy. At centimetre wavelengths the background radiation from the Milky Way and from distant radiogalaxies and quasars has become rather weak. The radiation from the atmosphere is also negligible so the sky should be very dark. What Arno Penzias and Bob Wilson, working at Bell Laboratories, found is that the sky glows faintly in the microwave band whichever direction you look in. They found this radiation by accident. They were trying to eliminate an unknown source of microwave noise in the Bell 20-foot horn reflector which had been built for communication via the Echo satellite. They wanted to use the reflector to measure the brightness of the radio source Cassiopeia accurately and also to measure the background radiation from the Milky Way, but the signals they were getting were hard to measure precisely because of the noise from the antenna. However no matter what they did they could not get rid of the noise. They chased some nesting pigeons out of the horn antenna and completely took it apart to clean it and remove what Penzias has referred to as 'a white dielectric substance', but still the noise persisted. Finally they concluded the noise must be radiation from the sky, coming equally from every direction. They did not immediately appreciate the significance of this until they talked to some physicists at Princeton University, who were themselves preparing an experiment to look for radiation left over from the fireball phase of a Big Bang universe. In 1978 Penzias and Wilson received the Nobel prize for this exciting discovery.

The brightness of the radiation is what would be expected from cold matter at three degrees above absolute zero (i.e. at –270 °C). The radiation has its peak brightness at a wavelength of about 2 millimetres and fades away quite sharply at wavelengths on either side of this. The smoothness of the background and the degree to which it looks the same in every direction are truly remarkable. When we looked out in the visible band at the galaxies we saw that they are spread out in every direction and that the distribution is fairly smooth apart from their tendency to collect in groups and clusters like our own Local Group of galaxies and the giant clusters in Virgo and Coma Berenices. But the sky of galaxies still looks decidedly patchy and irregular and the similarity of the distribution in different directions is no better than to a 10 or 20 per cent accuracy. The microwave background radiation on the other hand is isotropic (the same in every direction) to one part in a thousand or better. Now mathematicians had been making idealized 'models' of the universe which were perfectly isotropic, ever since 1917. In that year Albert Einstein had the audacity to suggest such a simple structure for the universe. Not many of these cosmologists can have genuinely expected the universe to really be isotropic like the mathematical models yet here was direct evidence that the universe really is very uniform and isotropic on the large scale. Efforts to explain the microwave background radiation as coming from galaxies soon ran up against the incredible smoothness of the background. It had to have an origin from the universe itself.

Let us travel back along the path that these microwave photons have taken to reach us. We find ourselves travelling onwards and onwards, past the most distant visible galaxy, on past the quasars, and back and back in time. Still the radiation seems smooth and uniform and we are nowhere near its source. We travel on past the epoch of the first generation of stars in the galaxies. Now the galaxies are nothing but clouds of gas falling together. The universe seems more crowded and the proto-galactic clouds of gas are much closer together than the galaxies were at the start of our journey. The background radiation is much more intense and hotter, and the wavelength of the peak intensity is now in the infrared. For it is not just the visible photons from galaxies that have their wavelength shifted towards the red by the universe's expansion. The microwave background photons have suffered the same fate and were originally shorter wavelength infrared photons. Now the clouds are touching and merging into each other. The gas begins to look more and more smooth in its distribution and, because we are travelling back in time and continue to see the expansion of the universe in reverse, ever denser. We can no longer see much sign of the irregularities in the gas which are later to form into galaxies. The temperature of the radiation is getting higher and higher, more than 1500 °C. The gas is heating up too and beginning to ionize. As the electrons are stripped from the atoms and move around freely they begin to scatter and absorb some of the photons. We are reaching our journey's end. Suddenly our photon is gobbled up by a free electron. The gas has become fully ionized and the universe is opaque to radiation. It is as if we were suddenly trapped inside a star. At this point the universe is one thousand times smaller than when we set off.

The fireball phase of the Big Bang

We have reached back in time to what is called the 'fireball' phase of the universe, because it resembles the fireball of an H-bomb explosion. The path of our photon now becomes a random zigzag to and fro as it is scattered, absorbed and re-emitted many times. As we travel back further in time the gas and radiation get hotter and hotter. They are now interacting with each other so rapidly that they share out any energy equally, so that the average photon and the average atomic particle carry the same energy. Eventually we reach back to a time when the gas is hot enough for nuclear reactions to be taking place. This is the stage when most of the helium and deuterium that we see today in the universe were made. We have already seen that most of the other elements are made in the interior of stars.

We travel on back in time. The matter of the universe is now in the form of a soup of the elementary particles of which atoms are built. Naturally there are protons and electrons, the basic particles that make up a hydrogen atom. But now we start to notice a growing proportion of antimatter, mirror image particles. For example the antiparticle of the electron is the positron, with a positive charge instead of a negative charge. Our photon has so much energy now that it might spontaneously turn into an electron–positron pair. Conversely, if a positron and an electron collide, they vanish in a puff of energy, namely a photon. This transformation of particles to photons and of photons to particles is an illustration of the equivalence and interchange-ability of mass and energy predicted by Albert Einstein from his Special Theory of Relativity in 1905. The famous equation $E = m c^2$, where m is the mass, c is the speed of light, and E is the energy, has had dramatic applications in our nuclear age.

Travelling on back in time, we come to a time when there were as many electrons as positrons in the dense cosmic soup of particles. On earth positrons are very rare and short-lived compared to the stable electrons. Earlier still we see large numbers of 'neutrinos', a strange weightless and unsociable particle which on earth hardly interacts with other matter at all. (The ghostly neutrino is created profusely in certain nuclear reactions in the sun: most of them escape from the sun and rush right through the earth as if there was nothing here at all. At midday they travel through our bodies from head to toe and at midnight from toe to head.) Conditions are getting so extreme in the dense fireball that even neutrinos begin to collide with other particles. The temperature of the universe is now a hundred thousand million degrees. We have reached back as far as we can reasonably extrapolate from the microwave background radiation reaching earth. If we extrapolate the expansion of the universe backwards in time for one hundredth of a second more we reach the infinite densities of the initial explosion, the Big Bang which gives this model of the universe its name. We have travelled back in time twenty thousand million years. Yet the whole fireball phase lasted only a million years, a short time on the cosmic scale.

What can we say about the first hundredth of a second of the history of the universe? The best analogy seems to be with the collapse of a star to form a black hole, except that time is reversed. As the temperature rises above a hundred thousand million degrees on our journey backwards in time, we enter the 'hadron' era, when the energy of a photon is sufficient to create a heavy particle, or hadron, and its antiparticle, spontaneously. Examples of hadrons that we have met are protons and neutrons. Others have exotic names like pi meson or lamda hyperon. Theoretical physicists do not agree about what the hadron era was like. One possibility is that it consisted of 'quark soup'. The 'quark' is the hypothetical elementary particle which comes in various 'flavours' and 'colours' and out of which the hadrons are built. If quarks rather than hadrons are the basic elementary particle then at an early enough time in the history of the Big Bang the universe would have consisted of a soup of electrons, positrons, neutrinos, photons, and quarks all sharing out the energy equally. As we travel on back in time we eventually come to a fundamental limit when general relativity breaks down because of quantum effects. This occurs when the temperature of the universe was a hundred million million million million million degrees and the time after the beginning of the explosion was one second divided by one with 43 noughts after it. We certainly have no idea how the universe was before this time.

Our journey back into the fireball went far beyond our brief of a voyage through the microwave landscape. It is certain that the photons of the microwave background come from very great distances and tell us something about the early history of the universe. The above account of the fireball is how contemporary theoreticians make sense of the microwave cosmic landscape. Even as I write I am aware that the latest measurements of the background from a balloon flown by a group at the University of California, Berkeley, do not quite fit in with what is expected on this 'fireball relic' picture. Science is always like this. There are loose ends hanging off the tapestry everywhere. The scientist pulls at them and sometimes they turn out to be some kind of mistake or misunderstanding. But occasionally the whole tapestry comes apart and has to be started again.

The molecular landscape

Our microwave journey falls naturally into two parts and we now return to earth to tackle the second mystery, life's complex molecules. Until recently astronomers thought that most of the matter in the universe was either atomic gas or in the ionized state that we have learnt is so common in the cosmic landscape. One per cent of the matter, the 'heavy' elements, was known to be in the form of solid grains of dust. But earth's molecular atmosphere, with its creatures built of complex molecules breathing it, was thought to be a rare anomaly. It has been understood for a long time, though, that the surface layers of cool red stars contain certain molecules like carbon monoxide (CO) and the cyanogen radical (CN). This can be seen by their aborption of their characteristic wavelengths from the star's visible light and is one of the ways cool red giant stars can be recognized and classified.

Microwave astronomy has changed this, for we now realize that about half the interstellar gas of the Milky Way is in the form of molecules. To understand why the microwave band is important for molecular astronomy, we have to think a little about what a molecule is and how it radiates. Take a simple molecule like carbon monoxide. In a carbon atom a cloud of electrons orbit round a nucleus of protons and neutrons and the same is true for an oxygen atom. The difference between the two atoms lies only in the number of protons, electrons, and neutrons in each. The carbon atom has 6 each of protons, electrons, and neutrons, whereas the oxygen atom has 8 of each. In the carbon monoxide molecule the two nuclei are immersed in a common cloud of electrons. The two electron clouds partially merge together.

Some electrons now orbit round both nuclei and it is these motions that hold the nuclei together through electric forces. The nuclei vibrate to and fro like weights on a spring and they also rotate round each other like a double star system.

If the molecule absorbs energy, either by colliding with another atom or molecule or by absorbing a photon, there are three ways it can get excited: the electrons can move to a more energetic orbit further out from the nucleus; the vibrations along the axis of the molecule can get more energetic; or the whole molecule can rotate faster. The excited molecule can now dispose of its excess energy by changing to a less energetic state and radiating a photon away. If the photon comes from a change in the energy of the electron orbits, it will be an ultraviolet photon. We have already seen the effect of absorption of such photons by clouds of molecular hydrogen on our third voyage. Changes in the vibrational energy result in the emission or absorption of infrared photons. Such photons are only just now beginning to be detected by astronomers and they only tend to be observable where the molecular gas is rather hot, for example when a cloud of molecules has been suddenly compressed. Finally, if the molecule changes from one state of rotational energy to another lower state, then in general a microwave photon is emitted. The photons have an energy characteristic of the particular change of energy and of the particular molecule.

Let us return to the molecules of interstellar space. The first rotational transition from a molecule in the cosmic landscape to be detected was from the hydroxyl radical (OH) at a wavelength of 18 centimetres in the radio range. This was discovered by Sandy Weinreb and Al Barrett in 1963 using the 84-foot radio telescope at the Lincoln Laboratory at the Massachusetts Institute of Technology. Five years later radio astronomers at Berkeley, Caliornia, detected ammonia and water vapour at a wavelength of around 1 centimetre and in the following year formaldehyde was detected by a group at the US National Radio Observatory (NRAO) in West Virginia. But the real breakthrough in the astronomy of molecular lines came when Arno Penzias and Bob Wilson of Bell Laboratories built a sensitive millimetre-wave receiver and started to use it on the NRAO 36-foot telescope at Kitt Peak, Arizona, in 1970. This telescope was designed to work down to wavelengths of 3 mm but the manufacturers managed to do even better and the telescope still works at a wavelength of 1 mm.

Fifteen new molecules were discovered during 1970–1, Penzias and Wilson being involved with six of them. Over thirty different molecules have been discovered in the interstellar gas with the 36-foot telescope alone. The most important of these molecules is carbon monoxide with its characteristic rotational wavelength of 2.6 millimetres. Many of the simple molecules made up of two or three atoms of carbon, nitrogen, and oxygen have now been seen in the cosmic landscape. Molecules with up to 11 atoms have been found, including that friend (or enemy) of mankind, alcohol. The discoverers of cosmic alcohol estimate that a scoop the size of a football pitch dragged through a dense interstellar cloud of molecules would hardly even yield one decent drink, and that the proof is approximately that of Milwaukee beer. The most important absentee from the list of molecules observed in the microwave band is the most abundant of all, molecular hydrogen. This is because the hydrogen molecule is too simple and symmetrical to undergo rotational transitions like other molecules.

Molecular clouds and the molecules of life

Let us select the characteristic rotational wavelength, 2.6 millimetres, of the carbon monoxide molecule, the most abundant in the Milky Way after hydrogen. We see a band of light along the half of the Milky Way nearest to the galactic centre in Sagittarius, stretching from Cygnus to Perseus. From the other half of the Milky Way we see only isolated patches of emission. This tells us already that the molecular gas in the Galaxy is more concentrated towards the centre than the stars or neutral atomic gas.

The carbon monoxide emission from the Milky Way is not a smooth band of light. It is broken up into patches of brighter emission. Many of these clouds of molecules turn out to be the same as the clouds of dust we saw radiating in the infrared and which made dark patches on the sky in the visible. The main constituent of these clouds is molecular hydrogen, with about 1 per cent in the form of dust grains and other molecules. The clouds are denser than the typical cloud of atomic hydrogen found by the 21-centimetre radio line of neutral hydrogen. Most are very cold, only ten degrees above absolute zero.

We travel towards the constellation of Orion, to the strong

infrared source in the Orion nebula. The carbon monoxide molecules are unusually hot at the position of the strongest infrared emission, probably being heated by collisions with warm dust grains which are absorbing the light from young stars and protostars embedded in the cloud. The extent of the surrounding molecular cloud is immense, covering hundreds of light-years and spreading over most of the Orion constellation. Next time you look up at Orion, try to imagine this vast cloud of molecular gas with its star-forming core near the centre of the Hunter's sword. The far-infrared source corresponds to a dense core of molecules and dust where star formation is going on. The Trapezium stars and the Orion nebula itself make up another older condensation within the overall complex. We find a similar picture when we explore the neighbourhood of other dense dust clouds in which star formation has recently happened, though most are rather far away so it is harder to sort out the details.

It is within the dense cores of these molecular clouds that the great variety of interstellar molecules, over forty different molecules to date, have been found. Still more would be found if we knew exactly what wavelengths to look at or if the wavelengths did not lie outside our observing windows. The chemistry of the formation of these molecules is complex. It seems that the molecular hydrogen forms on the surface of dust grains. Two hydrogen atoms stick to a grain and then migrate together to form a molecule which is then kicked off the grain by an ultraviolet photon. The grains are needed not only for the molecules to form on but also to shield the newly formed molecules from the bulk of the normal ultraviolet radiation from stars. The rest of the molecules apart from hydrogen seem to be formed by chemical reactions which take place within the gas, but still need the high density of dust grains to shield them from ultraviolet light.

What is so surprising and exciting about these molecules between the stars is that most of them are 'organic' molecules, that is molecules based around chains of carbon atoms. It is the ability of carbon to form itself into very long chains that allows life's complex molecules to exist. Life can be thought of as a wonderful demonstration of the versatility of the carbon atom's chemistry. The largest molecules found to date in the cosmic landscape have nine or eleven atoms: ethyl alcohol, dimethyl ether, cyanohexatriyne and cyano-octatetriyne. Is there anything to prevent us finding molecules still closer to those of living creatures, for example the amino acids? The simplest amino acid is glycine, with ten atoms, no larger than the molecules already seen. The main problem has been finding what the characteristic wavelengths for the rotational transitions are. Recently these characteristic wavelengths were measured for glycine in the laboratory. It was then possible to search for glycine at these wavelengths in the dense molecular clouds where so many other kinds of molecules have been found. Glycine was not detected but now that we know where to look, it may not be too long before we find it in the cosmic landscape. We saw on our visible voyage that certain stony meteorites contain complex organic molecules, including many different amino acids.

The abundance of large interstellar molecules discovered so far is extremely small, and for larger molecules may be smaller still. However it is not known for certain what form the most abundant heavy elements, carbon, nitrogen and oxygen, take in these dense molecular clouds. If the clouds have the normal abundance found elsewhere in our Galaxy, then we know that most of these elements are not in the form of simple molecules like carbon monoxide, methane, and ammonia. This leaves room for speculation that they could be locked up in very large molecules not yet observed. One bizarre speculation put forward recently by Fred Hoyle and Chandra Wickramasinghe is that the carbon is mainly in the form of cellulose, one of the basic ingredients of plants. They claim this could also explain the ten-micrometre absorption band which on our infrared voyage we attributed to silicates. The difficulty is to think of a chemical scenario which would turn almost all the interstellar carbon into cellulose. The most likely location for the carbon, nitrogen, and oxygen is probably in icy mantles on the dust grains.

Whether or not some form of life can exist in these molecular clouds the organic molecules we have found demonstrate that the first steps along the path to life are relatively easy to take and have been taken in all the clouds we see where stars are now being formed. The solar system therefore formed out of gas rich in organic molecules. However in most of the scenarios for the formation of the inner planets, all but the simplest of these molecules would have been destroyed by heat. Mars and the moon, for example, seem to be devoid of organic molecules. Only when the earth had cooled down and formed its primitive atmosphere of carbon dioxide, methane, water, and ammonia, could carbon's versatile chemistry again take over.

When we look at the microwave cosmic landscape and see the molecular clouds in all their richness, we are perhaps looking at places where life just failed to come into being. The ingredients are there, but the specialized environments of the liquid and solid state which make up earth's surface are lacking. As we look at our barren planetary neighbours and as we consider the disastrous effect for life on earth of small changes in the environment, the garden of earth seems rarer and more precious than ever. Yet hostile though most of the cosmic landscape would be to our bodies, we are one with it. The atoms of our bodies have floated in molecular clouds and have been through the furnace of a star's centre. In the form of the much simpler atoms of hydrogen and helium, we were present in the fireball and in some unknown form we experienced the Big Bang. It is our own personal landscape.

8
RETURN TO EARTH

How did life arise on earth?

We return from our voyages through the cosmic landscape to earth. How gentle and alive it seems after the cold violence of the realm of the galaxies. How did so rich a landscape as earth, teeming with life, come into being?

We have seen that apart from helium and deuterium, the elements of earth had their origin in nuclear reactions in stars, from which they were released by the dramatic events of the last days of the stars' lives: stellar hurricanes from red giants, the ejection of planetary nebulae, or the cataclysm of a supernova explosion. We saw the dense, dark clouds of dust and molecules from which a star like the sun would have been born and, in these clouds, the T Tauri stars which are on their way to being new suns. The aggregation of earth from smaller fragments and floccules, or its condensation out of a disc of gas, remains shrouded in mystery. We do not know for certain that other planets as big as Jupiter exist and we certainly do not know if other planets like earth are to be found outside the solar system.

We know that the cloud out of which the solar system formed is likely to have been rich in organic molecules, possible even in living things. The formation of earth and the subsequent heating up of its interior would almost certainly have broken these organic molecules down to methane, ammonia, carbon monoxide and dioxide, water, and hydrogen. The earth's primitive atmosphere, formed by the volcanic degassing of the interior, would probably have had this kind of composition. This atmosphere would have been bombarded by the sun's ultraviolet radiation, including wavelengths as short as 2000 Å since there was no ozone to absorb wavelengths longer than this, and by electric discharges from lightning. Laboratory simulations have shown that if such a mixture of gases is irradiated by ultraviolet radiation or subjected to electric discharges, an amazing mixture of complex organic molecules is produced, including many of the amino acids which are the building blocks of life. About half of the 20 amino acids that occur in proteins have been synthesized in this way. The stony meteorites called carbonaceous chondrites, like the Orgeuil and Murchison meteorites I mentioned in Chapter 2, show a similar complex mixture of organic molecules, including amino acids, and these are believed to have their origin in the kind of prebiotic synthesis which must have taken place in the early days of earth's history.

Some progress has also been made in synthesizing some of the more elaborate molecules out of which DNA, the genetic material, is constructed. Once they are formed these larger organic molecules may have accumulated in the oceans and then been concentrated through evaporation, for example in tide-pools. The step from these concentrations of organic molecules to simple living things like bacteria and algae is still a huge one and may have taken 1.5 thousand million years on earth. The oldest rocks that have been studied are about 3 thousand million years old and show evidence of photosynthesizing organisms in them. For comparison the age of the earth is 4.6 thousand million years.

Once these simple organisms like bacteria and algae had formed, they paved the way for the explosive evolution of multi-cellular organisms that we see now in the fossil record, by creating an oxygen-rich atmosphere. This explosive evolution occurred at the beginning of the geological period known as the Cambrian, about 600 million years ago. After another 200 million years the first animals with backbones and land plants appeared, and after a further 200 million years the long age of the dinosaurs began. Then 65 million years ago the mammals took over the earth and the primates emerged, culminating in the appearance of human beings a few million years ago.

The calamities that life has survived

During the 3 thousand million years that life has existed on earth there have been repeated calamities that cannot have failed to leave their mark on the evolution of species. Every 250 million years or so there is a major ice period, lasting about a million years. In each ice period there are several ice ages, each separated by about 250 000 years and lasting about 50 000 years. The most recent ice age covered much of North America and Europe with a thick layer of ice and ended only 11 000 years ago. This is also the furthest back that we can find evidence that human beings engaged in agriculture. Our emergence from the last ice age marks the dawn of civilization.

There are two possible explanations of this regular succession of ice periods. One is that variations in the earth's distance from the sun and other wobbles in its motion could lead to long time-scale variations in the average temperature of the earth. The earth's climate depends very sensitively on this average temperature and a drop of only a few degrees is enough to trigger an ice age. The grim prediction of this theory is that the next ice age is already beginning. It will take a few thousand years to engulf us and then will last another 50 000 years. Before we start to get too gloomy about this, we should remember that all the creatures and plants we see today survived the last ice age without any help from modern science.

A second theory for the ice periods is that they correspond to the passage of the solar system through the spiral arms of our Galaxy. We saw on our infrared voyage that in spiral galaxies a spiral-shaped wave rotates round the galaxy compressing any gas clouds it encounters and triggering some of them into forming stars. The sun, with the planets in tow, orbits the Galaxy about once every 200 million years and should run into a spiral arm twice during that time. This is of the same order of magnitude as the interval between ice periods. When the solar system is passing through a spiral arm it has a much greater chance of running into one of these clouds of gas and dust. What would the consequence of this be? You might think that because these clouds blot out the light of the stars behind them, the sun would be extinguished. However you have to look through about a light-year of the cloud to get some noticeable dimming of visible light, so the sun, only 8 light-minutes away from earth, would not look any dimmer. Astronomy might get more difficult, though, since the visible light from the stars and galaxies might be dimmed by a factor of 10 or more.

The main consequence of passing through a cloud would seem to be a big increase in the amount of dust falling on the earth. It is this pollution that could modify earth's climate. The solar system would spend about 50 000 years if any one cloud and would travel in and out of several clouds while crossing the spiral arm. This fits in well with the duration and frequency of ice ages during an ice period.

Whatever the cause of ice ages, life on earth has survived many such crises, and will presumably survive many more in the future. Still, the next passage of the sun through a spiral arm may be an anxious era for the astronomers of the future. Quite apart from watching out for the dust clouds ahead of us which might plunge us into an ice age, they will be looking out on either side of our path too for red giants and supergiants. If one of these should be a star heavier than about 5 suns and near the end of its short life, we could find ourselves too close to a supernova explosion for comfort. Because the lifetime of such stars is short compared with the time it takes the sun to go round the Galaxy, we only encounter them near their birthplace, in a spiral arm. Supernovae are not very common, though, so it is not very likely that we would find ourselves close enough to one (i.e. within a few light-years) for its blast of cosmic rays to wipe out instantly ourselves or other species. However it is quite likely that we will find ourselves within, say, 30 light-years of a supernova explosion during our next spiral arm passage. This is near enough for the earth's protective ozone layer to be severely depleted by the cosmic ray blast, letting through the dangerous shorter ultraviolet wavelengths of the sun. The near-by supernovae of the past must have affected the evolution of species and will be a crisis for life again in the future.

A rather similar crisis occurs due to a strange phenomenon of the earth itself. By studying the magnetism of ancient lava deposits on the ocean beds, scientists have found that every so often the earth's magnetic field drops to zero and changes over to the opposite polarity (earth's north and south magnetic poles change places). The earth's magnetic field is believed to be generated by convective motions of the earth's liquid iron core (it is these same convective motions that slowly but steadily shift the continents over the surface of the earth). But the reason for the reversals of magnetic polarity are not understood. The earth's magnetic field remains close to zero for about 10 000 years and during this time cosmic rays from the sun would directly bombard earth's atmosphere, instead of being deflected away or trapped in the radiation belts as most are at present. The result, as with a near-by supernova, could be a reduction in the ozone layer, letting through the sun's shorter ultraviolet wavelengths. A massive solar flare might completely remove the ozone layer, so that exposure to sunlight would be fatal for many creatures, including man.

How common are these magnetic field reversals? The last one was 600 thousand years ago, and there have been periods of a million years without change. At other times the magnetic field appears to have flipped back and forth every 200 thousand years before settling down again. Life on earth has survived this crisis hundreds of times and even mankind has survived it several times. Either the consequences of a magnetic field reversal are not as severe as feared or mankind lived in a life style that did not expose him much to sunlight during the crisis millenia.

Most of these potential calamities for man and for life on earth occur only over a very long time-scale, a million or a hundred million years. On an even longer time-scale we can look forward to the catastrophic effect for the earth of the sun's embarking on its red-giant phase in about five thousand million years' time. The earth will probably be swallowed up inside the sun's huge surface.

We can also speculate on the future of the universe itself. At present it looks set to keep on expanding for ever, because the gravitational attraction between the galaxies is not sufficient to halt the universe's expansion. However our estimates of the amount of matter in the universe are very uncertain because there could be large amounts of matter in an invisible form (black holes or intergalactic asteroids, for example). If we have under-estimated the amount of matter around in the universe by a factor of ten, then the expansion will eventually be halted and the universe will collapse back to a state of infinite density. This fireball implosion, should it occur, is likely to be at least 100 thousand million years away. If this is not our fate and the universe is to keep on expanding, then the galaxies will eventually fade out as their stars convert all the available hydrogen to heavy elements. We saw on our visible voyage that stars much heavier than the sun are profligate with their nuclear fuel and do not live long. The longest lived stars are those of lowest mass: for example, stars of one-tenth the mass of the sun will last a million million years.

The possible futures for mankind

Now even a hundred million years is a very significant time on the evolutionary time-scale. It is a bit less than the time the dinosaurs ruled earth and it is a bit more than the time the mammals have been supreme. Over this time-scale it is meaningless to speculate about the future of the human race, since we may be unimaginably improved as a species or we may have become extinct.

Let us choose a more reasonable time-scale and try to imagine what the future holds. If we choose a hundred years, say, then it is easy to imagine that things will not be very different from the present. There will be anxieties about food, energy, resources, population, pollution, just as at present. Things may have got slightly better, or slightly worse, but they are hardly likely to have reached utopia or doomsday. Instead let's take a more interesting time-scale, a thousand years.

One obvious possibility a thousand years from now is that advanced technology has collapsed due to exhaustion of resources and energy. Mankind has reverted to a simpler, rural life, and the technological breakthroughs of the past century have been forgotten. This kind of collapse has happened before, for example after the fall of the Roman Empire. And we should remind ourselves that although such a collapse of technology would affect the lives of a far larger proportion of the world's population today than the fall of Rome did, it would probably still not be a majority who were profoundly affected.

It may be a consequence of some of the decisions that we have already taken, or rather, that have been taken for us, that this option of a reversion to a simple, rural life is no longer open to us. We are going to need scientists and science for thousands or even tens of thousands of years simply to look after the nuclear wastes and abandoned nuclear power stations. We can imagine a scene far in the future when science and technology have long ago collapsed, but a brotherhood of initiates cling to the remnants of knowledge about atomic physics. How powerful they would be, like the astronomer–astrologers of the Maya or the ancient Chinese, talking a mixture of sense and nonsense, nobody able to undertake any project without consulting them. The nuclear waste dumps and power stations, buried beneath pyramids of concrete, would have the same mystical and terrible associations in the popular mind of the future as the pyramids of ancient Egypt or of the Maya.

Another possible scenario for the future, and a totally new one on the historical stage, is that we could destroy ourselves utterly with our absurd arsenal of nuclear weapons. Perhaps that is what happens to advanced civilizations and is the reason why the cosmic landscape seems so empty apart from ourselves. If so, our existence seems like a very cruel joke. That so much pain and effort should go into developing our culture to its present level only to be wasted in an instant's folly.

A third scenario is that our technological development continues at the present pace for the next thousand years, opening up possibilities that seem beyond the imagination at present. We might learn how to fly the solar system and be able to keep it out of trouble from Galactic dust clouds and supernovae. The speed-of-light barrier could melt away, making space travel and interstellar communication commonplace. Our life in the cosmic landscape may only just be beginning.

Is there anybody out there?

What are the chances of encountering other intelligent life in the universe? Speculation about life on other worlds goes back at least 2000 years. In our own times these speculations are being taken rather seriously and there are men and women who earn their living trying to communicate with other worlds (e.g. under NASA's SETI (Search for Extra-Terrestrial Intelligence) programme). Attempts are made to calculate how many contemporary communicating civilizations there are likely to be in our Galaxy, assuming that they need conditions similar to earth in order to have evolved. The answers range from 0.000 000 1 to a thousand million, reflecting the hopelessness of the calculation in our present state of knowledge. The first, pessimistic figure means that our own civilization is a freak occurrence. The second, optimistic figure means that a reasonably near-by star should have a planet with an advanced civilization on it.

The calculations are so uncertain because there are four totally unknown quantities which have to be multiplied together with some other quantities that are reasonably well known. For example we know how many stars there are in our Galaxy similar to the sun (about 30 thousand million with masses between one-half and one and a half times that of the sun) and we know how many of these are single stars (10–15 per cent). Then we come to the first complete unknown: how many of these single stars like the sun have earth-sized planets at the right distance from them? If the solar system planets arose from some rare, chance event like the encounter of the young sun with a dense protostellar cloud, the probability of a star like the sun having planets like the earth could be close to zero. On the other hand if the solar system planets are a natural by-product of the process by which the sun formed, this probability could be high. Most planetary theorists favour the latter kind of model, but this is partly a matter of individual psychology or taste. It is more interesting and rewarding to attempt to explain the solar system by a universal type of mechanism which would apply to other stars too. The matter can only be settled by finding other planetary systems and that will be difficult.

The next step in the calculation, the probability of the formation of organic molecules from the material of the primitive atmosphere and ocean, seems easier. The laboratory experiments on simulated primitive atmospheres, the organic molecules in the carbonaceous chondrite meteorites, and the rich array of organic molecules found in the dense clouds from which stars are forming, all point to the ease with which organic molecules form in the cosmic landscape.

Now we come to the second totally unknown factor, the probability that these molecules organize themselves into life. In our present state of ignorance the probability could be anywhere from close to zero to one. Our existence shows that it did happen once, but the disappointing results of the *Viking* mission to Mars is a serious blow to the optimists.

The third unknown factor is whether when life arises it is likely to evolve towards intelligent creatures and whether they will ultimately develop advanced technology so that we can communicate with them. The wonderful variety of living things that we see on earth, the exuberance with which nature has filled every ecological niche with some creature or plant, make it hard to doubt that any life will ultimately evolve a creature intelligent enough to appreciate nature itself and to travel in imagination through the cosmic landscape.

Finally there is the question of how long a technological civilization lasts. When we hear a signal from the cosmic landscape, will it just be a wreck marker-buoy, the last message of a civilization close to destruction? Do we have time to create such a message ourselves before we destroy earth?

What would it be like encounter them?

Supposing the optimistic calculations are correct, what are we likely to encounter out there? It seems exceedingly unlikely that our first encounter would be with people at the same stage of development as ourselves. Our forty years with radio telescopes follow three thousand million years of evolution. Even if others developed in an almost identical way they are unlikely to reach this stage at the same date or even within a thousand or a million years. There are stars a thousand million years older than the sun just as suitable for the evolution of life. When we think of the voyages of discovery of the European Renaissance, the navigators and travellers found peoples at vastly different stages of development, even though these peoples all had a common origin. We who have only just begun to see the cosmic landscape are likely to be the undeveloped primitive in any interstellar conversation. It is also inevitable that with their thousands or millions of years of technology they will discover us before we discover them. Our earliest radio broadcasts at frequencies high enough to penetrate the ionosphere may even now be being detected by creatures with unimaginably advanced technology, who are now deciding whether to ignore us, colonize us or look after us. It seems to me that if such creatures exist relatively near earth we are in a very vulnerable state. Perhaps our days as an independent planet are numbered. In sending out signals deliberately beamed at near-by stars we are giving a hostage to fortune. Would the Indians of the Americas have sent such a signal to Europe before 1492 if they had been able to? We have to hope that civilizations much more advanced than ourselves are less aggressive, less selfish, and less violent than we are at the present era. Perhaps they have to be to survive the discovery of nuclear physics. But perhaps they have to be selfish to survive.

If science and technology continue to develop on earth for thousands or millions of years, it is clear that we shall be taking a great interest in the question of other habitable planets, so that we can emigrate either because of some kind of cosmic calamity here (change of climate, reversal of magnetic field, near-by supernova) or just for fun. Others may take a similar interest in earth now and might find it very underpopulated and rich in natural resources compared with their own planet.

Human beings have a natural hunger to know if there is other intelligent life in the cosmic landscape, even if only so that we can wave to each other across the gulf of interstellar space and feel less alone in the immensity of space and time. Yet the most likely circumstances of such an encounter, with creatures far more developed than ourselves, seem fraught with insecurity and danger. An encounter with creatures like ourselves does not seem an attractive idea either, with both planets changing their inflated military machines over from 'defence' weapons directed against other nations to 'defence' weapons directed against other solar systems.

The ideal encounter would be with some civilization slightly more advanced than ourselves, mature, gentle, with no interest in colonizing us, who could help us sort out our problems. This hunger for a cosmic parent or psychiatrist will get worse if things keep on going so badly on earth. But is it likely that such people exist? Why are we here after all? There seems to be no reason, except as a product of the programme of certain molecules to replicate themselves. The molecules made us, but they did not necessarily have in mind coping with the nuclear age. Was there no advantage to the molecules in making us more mature?

Yet we do not have to be ruled by the molecules, for our cultural evolution enables us to transcend their limitations. And it is our culture that would help us bear slavery if that were to become our fate. Like the ancient Greeks under Roman rule we would intellectually and spiritually colonize our conquerers, transforming them into Byzantines.

In the face of these apocalyptic visions of the future and of the vague menace of civilizations a thousand or a thousand million years more advanced than ourselves, we turn again to earth, to the secure garden of the terrestrial landscape. We do not do so in quite the same frame of mind as the medieval artists. Our *Vision of Judgement* is not peopled by demons like the landscape of Hieronymous Bosch, though the creatures that alarm us might look just as fantastical. And the garden we return to is not quite the tame, ornamental one of the *Lady with the Unicorn* tapestries in Paris. A better parallel is with Leonardo da Vinci, with his interest in astronomy, his scientific desire to describe and understand nature, and his obsession with the idea of the Deluge, a product of the apocalyptic speculations current around 1500.

We have seen that we can experience the cosmic landscape in all its aspects by an imaginative extension of our senses. Our bodies can feel the infrared radiation of the sun and our skin responds to the ultraviolet by getting tanned. The X- and gamma rays of the cosmic landscape, mainly from the surface of our radioactive earth, have had a profound effect on our evolution from primitive forms of life through the mutations they cause. The human body radiates mostly in the infrared band, but it also radiates weakly at microwave and even at radio wavelengths. We can think of the astronomer's diverse telescopes, and the strange detectors with which he records radio, microwave, infrared, ultraviolet, and X- and gamma radiation, as an extension of our eyes. In the first chapter we called this the all-frequency-light-machine.

With these new eyes we saw a series of landscapes, or rather views of the same landscape. In the visible band we saw beyond the landscape of planets and stars to giant star systems like the Milky Way, the galaxies, spread out in every direction as far as the eye could see. In the radio we saw along the Milky Way clouds of hot ionized gas around hot stars and by contrast very cold clouds of gas where stars have not yet formed. We saw the astonishingly regular flashing of the pulsars, hinting at the dramatic death of stars as supernovae and neutron stars. Beyond the Milky Way we saw the violent activity of radio galaxies and we discovered the awe-inspiring power of the quasars, power equivalent to a thousand Milky Ways from a volume no bigger than the solar system. In the ultraviolet band we were surprised by the million-degree gas of the sun's corona. The hot, young luminous stars and the hot, dying white dwarfs blazed out. We saw that many galaxies have miniature quasars in their very heart. With X-rays the dead stars were brought to life, white dwarfs, neutron stars, and the sinister black holes, lit up by gas from their living companions pouring onto them. We saw the full drama of a solar flare and the other active sources, quasars and Seyfert galaxies, raged too. The giant clusters of galaxies glowed with gas stripped from their member galaxies. In the infrared band we saw the planets and asteroids of the solar system at their strongest. Stars shrouded in dust swam into view, both stars being born and stars close to death in their red-giant phase. We saw the clouds where stars are forming and the cold clouds whose turn is yet to come. We saw galaxies with quasars in their centre shrouded in dust. In the microwave band we saw the cosmic whisper that was all that remained of the Big Bang and we travelled back along the photon's path to the fireball itself. The world of molecules lit up and we found everywhere the organic molecules that are the first steps towards life itself.

And this landscape is our own personal landscape. The atoms of our body were in the cold clouds and in the hot centres of stars. Earlier they floated on the fireball of light that was the Big Bang. Yet looming larger still in the foreground is the landscape of earth, more important to us than all the rest of the universe. We have to care for this precious and rare environment that made us and that we are now beginning to be capable of destroying. There may be no other earths in the Milky Way. We probably do not have any contemporaries out there. Earth and its life are a marvel and mankind, with its culture, its sciences, and its arts, is the greatest marvel of all. The barren landscapes of the moon and Mars appear more familiar and real to us than the remoter parts of the cosmic landscape because they seem to have been seen through our own eyes and have acquired a human dimension. I have to admit that it is harder to feel the same way about the landscape seen with the telescope, the landscape of the invisible wavelengths, though the whole aim of this book has been to bring that landscape closer.

Is this cosmic landscape the only one we could have lived in? Are there other possible universes, with other histories, in which we could have existed? Our existence seems bound up with so many aspects of the cosmic landscape, perhaps all aspects if we could only see the connection. This point of view has been elevated by some cosmologists to a law, the anthropic principle: things are as they are because if they were not we would not be here. Several strange coincidences in the ratios between physical constants can be explained in this way, for all the different physical forces, gravitational, electromagnetic, nuclear, have played their part in our evolution. If gravitation had been a bit stronger or a bit weaker, stars like the sun or planets like the earth might not have formed and we would not be here. We see that the main scientific technique of the past four hundred years since the time of Galileo, that of examining a particular physical process in isolation from all others, is not suitable when we come to look at the universe as a whole. There is only one universe, the parts of it are not separate, and it is not a laboratory with many different experiments bubbling away. It is a complete and interconnected unity and we are an integral part of it.

Travelling through the cosmic landscape is a wonderful experience and we do not have to let the vast distances and times intimidate us. It is absurd to be depressed by the thought of the sun's extinction 5 thousand million years hence, or the universe's possible collapse back to infinite density in a hundred thousand million years. It is we, a mere bag of molecules scurrying round on a grain of interstellar dust, who have made this voyage and given the universe a consciousness of itself. The universe might have existed without us, but it would not have become a landscape.

GLOSSARY

Ångstrom (Å) a unit of wavelength, a ten-millionth of a millimetre.

Big Bang the explosion of the universe from a state of virtually infinite density which seems to have occurred about twenty thousand million years ago.

black hole matter so condensed that light (and astronauts) cannot escape from it. The sun would have to be compressed to a diameter of six kilometres or less, and the earth to three millimetres, to become a black hole.

blue star a star with a surface temperature in the range 10 000–30 000 °C looks blue to the eye. A star three to thirty times as heavy as the sun looks blue while it is fusing hydrogen·to helium.

Cepheid variable a kind of star, of which δ Cephei is the prototype, which pulsates and varies its light output regularly over a period of hours to days. The more luminous the star the longer the period of variation.

chromosphere thin layer of hot gas just outside the visible surface of the sun, named after its red appearance during eclipses.

constellation group of bright stars near together on the sky (not necessarily near in space) which make up a recognizable pattern.

corona zone of gas at a million degrees centigrade which surrounds the sun and extends out to a distance from the surface several times the sun's radius. Seen as an irregular halo round the sun during a total eclipse.

cosmic rays charged atomic particles (electrons and atomic nuclei) arriving at the earth at speeds very close to that of light, and therefore with very great energy, believed to be accelerated in pulsars (q.v.) or during supernova explosions (q.v.).

Doppler shift a source of light (or sound) has its wavelength or frequency shifted if it is moving towards us or away from us. Visible wavelengths are shifted towards the red (long-wavelength) end of the spectrum (q.v.) if the source is receding, and towards the blue if it is approaching.

dust most of the atoms heavier than helium in between the stars are in the form of very small grains of dust, about a tenth of a micrometre in size. The most common types of grain are probably graphite, silicon carbide, and silicates coated with ice.

ecliptic the planets all move round the sun in one plane, the ecliptic plane, so they seem to trace out a line round the sky, the ecliptic.

electron   light atomic particle with negative electric charge.

free–free radiation   radiation from an ionized gas (q.v.) due to electrons moving in the electric field of protons or other ions (q.v.).

galaxy   giant star system like the Milky Way, of mass ranging from a hundred million to a million million times the mass of the sun. The three main types of galaxy are elliptical, spiral, or irregular in appearance. Most galaxies have a central condensation of stars and gas, called the nucleus. About 1 per cent of galaxies show signs of violent activity in their nuclei (rapid motions, radio, or X-ray emission).

hertz   unit of frequency, one cycle per second.

H II region   cloud of ionized (q.v.) hydrogen surrounding a luminous blue star. Found along the spiral arms of spiral galaxies.

ionized gas   gas in which most of the atoms have had some of their electrons stripped off either by collisions between atoms if the gas is hot enough or by energetic ultraviolet or X-ray photons (ionizing radiation). The upper levels of the earth's atmosphere are ionized (the ionosphere).

ion   atom which has lost one or more electrons, leaving it with a positive electric charge. The ion of the hydrogen atom is the proton.

Lyman α-emission   ultraviolet radiation of wavelength 1216 Å, due to transition of the electron in a hydrogen atom from its second lowest permitted energy state to its lowest permitted energy state. Transitions from other states to this latter state give other wavelengths of the Lyman series. These wavelengths are characteristic of an ionized gas (q.v.) of hydrogen.

Magellanic clouds   two fuzzy or nebulous patches of light in the southern sky, due to small irregular galaxies (see galaxy) which orbit round our own Milky Way galaxy. Noticed by the navigator Magellan during his fatal voyage of 1521–2.

magnetosphere   region of space surrounding the earth (not actually spherical) controlled by the earth's magnetic field. The latter shields earth from the direct impact of the solar wind (q.v.).

Messier 1–103   list of fuzzy or nebulous objects compiled by the eighteenth-century French comet watcher, Charles Messier.

neutron   heavy atomic particle with no electric charge.

neutron star   tiny kind of dead star only tens of kilometres in diameter, in which the matter is in the form of neutrons crushed together till they touch. Neutron stars are formed in supernova explosions (q.v.) and are responsible for pulsars (q.v.).

non-thermal radiation   radiation that does not come from a hot body (see thermal radiation) but from some other process, for example from relativistic particles (q.v.), like synchrotron radiation (q.v.).

photon   particle of light. Light can be thought of either as waves or as a stream of particles. A photon carries a definite amount of energy, proportional to the frequency of the light, so the high frequency ultraviolet and X-ray photons carry more energy and have greater penetrating power than the lower frequency radio and infrared photons.

proton   heavy atomic particle with positive electric charge.

pulsar   pulsating radio source due to a rotating neutron star (q.v.).

quasar   star-like source of optical and ultraviolet radiation in the nucleus of an active galaxy (see galaxy), of such luminosity that the starlight from the galaxy cannot be seen.

red giant   red star (q.v.) much larger than the sun. Stars become red giants when they have exhausted the hydrogen in their cores.

redshift   the light from all but the nearest galaxies is shifted in wavelength towards longer wavelengths (in the visible band, this means towards the red end of the spectrum, q.v.). This is believed to be due to the Doppler shift (q.v.) and shows that the galaxies are receding from us — the expanding universe.

red star   a star with a relatively cool surface (2000–3000 °C) looks red to the eye. It radiates most of its energy at infrared wavelengths.

relativistic particles   atomic particles, e.g. electrons or protons, moving at speeds close to that of light (thirty thousand kilometres a second). Such particles often radiate non-thermal radiation (q.v.).

Roche point   if two stars are orbiting round each other very close together and one starts to grow in size, it becomes more and more elongated towards its companion, eventually acquiring a cone-shaped probuberance (the Roche lobe). Any further attempt to grow results in the star overflowing material through the apex of the cone (the Roche point) towards its companion.

scattering   when light is reflected in a random direction, for example by a grain of dust (q.v.), it is said to be scattered.

Seyfert galaxies   a class of active galaxies (see galaxy) in which rapid motions can be seen in the nucleus and with excess ultraviolet radiation compared to a normal galaxy, due to the presence of a compact source like a miniature quasar (q.v.).

solar wind   outflow of gas and relativistic particles (q.v.) from the surface of the sun, which impinges against the earth's magnetosphere (q.v.).

spectrum   the spread of wavelengths from a source of light, for example by a prism. The rainbow is the sun's visible spectrum.

supernova   explosion at the end of the life of a massive star in which much of the star's surface layers are thrown off and the star's core implodes to form a neutron star (q.v.).

synchrotron radiation   radiation from relativistic particles (q.v.), especially electrons, spiralling in a magnetic field.

thermal radiation   the radiation from a hot or warm perfectly absorbing body has a characteristic spread of wavelengths in which the brightest frequency is proportional to the temperature of the body, and the brightness then fades away very rapidly towards higher frequencies and less rapidly towards lower frequencies. The technical term is black-body radiation.

white dwarf small type of dying star thousands of kilometres in diameter, with a surface temperature of fifty to a hundred thousand degrees centigrade, in the process of cooling off towards a dark, cold, dead state.

zodiac the line on the sky marking the zodiacal or ecliptic plane (q.v.) in which the planets and sun move, marked out by the twelve constellations of the zodiac.

zodiacal light band of light along the zodiac due to scattering of starlight by small particles of dust (q.v.) spread through the ecliptic or zodiacal plane.

INDEX